Henry S. Williams

Key to Steam Engineering

Henry S. Williams

Key to Steam Engineering

ISBN/EAN: 9783743467255

Manufactured in Europe, USA, Canada, Australia, Japa

Cover: Foto ©berggeist007 / pixelio.de

Manufactured and distributed by brebook publishing software (www.brebook.com)

Henry S. Williams

Key to Steam Engineering

KEY

TO

STEAM ENGINEERING.

• • •

EMBRACING

IMPORTANT QUESTIONS ANSWERED

CONCERNING THE

STEAM ENGINE AND BOILER,

BY A PRACTICAL ENGINEER OF LARGE EXPERIENCE.

——

THE COMBUSTION OF COAL

AND ITS

ECONOMICAL USE CONSIDERED.

——

VAPORIZATION, EBULLITION, EVAPORATION,

EXPANSION AND CONDENSATION OF WATER;

ETC., ETC.

——

THIRD EDITION, REVISED AND ENLARGED.

PUBLISHED BY

HENRY S. WILLIAMS,

65 FEDERAL ST., BOSTON, MASS.

1892.

E. B. Stillings & Co., Printers.

CONTENTS.

PART I.

IMPORTANT QUESTIONS ANSWERED
CONCERNING THE
Steam Engine and Boiler.

PART II.

COMBUSTION OF COAL.

PART III.

APPENDIX.

ELECTRICITY.

PREFACE.

THERE have been many works published on Steam and the Steam Engine, which, although treating the subject in a comprehensive and scientific manner, have, nevertheless, failed to accomplish that which was intended (the education of the engineer), simply because the terms used by the college-educated writers, and scientific character of these books, are beyond the comprehension of the great majority of the men who are expected to profit by their contents.

In this little treatise all terms and tables not easily understood have been avoided, and in the simplest language possible the writer has briefly answered such questions as are likely to arise in the mind of an intelligent and practical engineer. using figures when necessary in their simplest form, and only such illustrations as will be readily understood by any one with a common school education.

Trusting that our efforts will be appreciated by those who desire to be qualified for the responsible position of an Engineer, but who find their opportunities limited for obtaining the requisite knowledge, this little Hand-Book is respectfully submitted by the Author.

PART I

SECTION I.

AIR.

Question. What three elements of Nature must we understand to successfully master steam engineering?

Answer. Air, water and fuel.

Q. What is air composed of?

A. Air is composed principally of the three gases — nitrogen, oxygen and carbonic acid gas, in the following proportions: nitrogen, four parts; oxygen, one part, with a slight admixture of carbonic acid gas.

Q. Which is the most important of these gases?

A. Oxygen is the most important, for to its agency are owing the existence of animal life, the maintenance of combustion, etc.

Q. What is the atmospheric pressure or weight?

A. The atmosphere has a pressure or weight of just 14.7 ($14\frac{7}{10}$) lbs. per square inch at the sea level. The higher we ascend in the air the less the pressure becomes.

Q. What causes the water to rise in a pump?

A. It is the displacement of air in the pump that causes the water to rise; the same may be said in reference to the siphon.

Q. What is a vacuum?

A. A space void of air.

Q. If we had a vacuum formed in a vessel, and were to suddenly open a communication between it and the outside air, how fast would the air rush in?

A. The air at its maximum density (14.7) is found to rush into a vacuum at the rate of from 1,300 to 1,400 feet per second.

Q. Which is the best conductor of sound. damp air or dry air?

A. Damp air is said to be the best conductor of sound.

Q. Is air a good conductor of heat and cold?

A. We heat or cool a room by the circulation of air; but air that is confined, or dead air, is a good non-conductor of either heat or cold.

Q. Does water contain any air?

A. There is about two per cent of air in ordinary fresh water; salt water contains less.

Q. What is the weight of air?

A. 13.817 ($13\frac{817}{1000}$) cubic feet of dry air at the sea level, with a temperature of 60° Fahr., weighs just one pound.

Q. Will an ordinary air-pump remove all the air from a perfectly tight vessel?

A. It will not remove quite all the air.

Q. Can a vessel be filled with either water or steam without first leaving some means for the air to escape?

A. The air must have a chance to escape as the vessel fills.

Q. In what part of the vessel must the vent be left for the air to escape when it is being filled?

A. At the top or highest point.

Q. What effect on the density of air does the compressing of it have?

A. If a volume of air were compressed into one-half of its space its density would be doubled, and if compressed into one-fourth of its space its density would be increased four-fold, and so on.

Q. How much does air expand with heat?

A. Air, at a constant pressure, expands $\frac{1}{461}$ of its volume for each (Fahr.) degree of heat communicated above zero.

Q. If air is kept at a constant density, how many (Fahr.) degrees of heat will it take to double itself in volume?

A. Just 480°.

SECTION II.

WATER.

Q. Of what is water composed?

A. Water is composed of oxygen and hydrogen.

Q. Is water compressible?

A. Water is compressible and is perfectly elastic, but the change is so minute as to have no practical consequence.

Q. How much pressure would be exerted at the bottom or base of a column of fresh water 27.71 ($27\frac{71}{100}$) inches high?

A. At a temperature of 62° Fahr., just one pound pressure; and a column of fresh water at the same temperature 33.947 ($33\frac{947}{1000}$) feet high would have a pressure of one atmosphere or 14.7 lbs. per square inch at the base.

Q. How high would a column of fresh water have to be to have a pressure of 15 lbs. per square inch at the base?

A. A column of fresh water 34 feet high, at a temperature of 60° Fahr.

Q. How much does water expand in freezing?

A. It is said that water in freezing expands about $\frac{1}{12}$, or .083 of its volume.

Q. What is the pressure at the base of a column of fresh water, at a temperature of 62° Fahr., one foot high?

A. Just .434 ($\frac{434}{1000}$) of a pound.

Q. What is the comparative weight of water?

A. Water is just 13.6 ($13\frac{6}{10}$) times lighter than mercury, and 815 times heavier than air at the sea level with a mean temperature (56° Fahr.)

Q. What is the weight of ice and snow?

A. One cubic foot of ice at 32° Fahr. weighs just 57.5 ($57\frac{5}{10}$) lbs., while one cubic foot of snow freshly fallen weighs 5.2 ($5\frac{2}{10}$) lbs. and has twelve times the bulk of water.

Q. What is the boiling point of fresh water?

A. Fresh water would boil in a perfect vacuum at a temperature of 72° Fahr., in the open air at the sea level at 212° Fahr., and under a pressure of 15 lbs. per square inch at a temperature of 234° Fahr.

Q. When is water at its greatest density?

A. Water is the heaviest, or at its greatest density, at a temperature of about 39° Fahr., or 4° Cent.; at this point it will expand either with the heat or the cold.

Q. If the water expands both ways from the above temperature, then it must be evident that there is a point at either side of this temperature where the water has the same weight. What are these temperatures?

A. Water at a temperature of 32° Fahr. has the same weight as water at 47° Fahr.

Q. What is the weight of a cubic foot of fresh water at a temperature of 60° Fahr.?

A. Just 62.37 ($62\frac{37}{100}$) lbs.

. What is the weight of a U. S. standard gallon of fresh water at a temperature of 60° Fahr.?

A. About 8.33 ($8\frac{33}{100}$) lbs., consequently there are about $7\frac{1}{2}$ gallons per cubic foot of water.

Q. What is the weight of an Imperial gallon of fresh water at a temperature of 62° Fahr.?

A. The Imperial gallon of fresh water weighs just 10 lbs., and it contains 277.274 ($277\frac{274}{1000}$) cubic inches.

Q. How many cubic inches does the U. S. standard gallon contain?

A. Just 231 cubic inches, or 294 cylindrical inches.

Q. How many cubic inches does a pound of fresh water contain at the mean temperature (56° Fahr.)?

A. About 28 cubic inches, or 35 cylindrical inches.

Q. How many Imperial gallons per cubic foot of water?

A. Just 6.232 ($6\frac{232}{1000}$) Imperial gallons.

Q. How much will water expand in rising from 60°
Fahr. to 212° (boiling point)?

A. About 2¼ per cent in volume.

Q. How many volumes of steam at 212° (when it first
rises from water) will one volume of water make?

A. About 1,700 volumes of steam.

Q. How many volumes of steam at 10 lbs. pressure
(240° Fahr.) will one volume of water make?

A. About 1,040 volumes of steam at 10 lbs., and about
765 volumes at 20 lbs. pressure (about 260° Fahr.), and
so on.

Q. About what is the average amount of salts contained in seas of the globe?

A. It is said that the average amount of salts contained in seas of the globe is about 3½ per cent; that is.
if a box 6 feet deep were to be filled with the average sea
water and evaporated or boiled away, there would remain
about 2 inches of salt at the bottom.

Q. What portion of the earth's surface does the sea
occupy?

A. About ⅗ of the earth's surface.

Q. What is the amount of curvature of one mile of
the ocean's surface?

A. About 2.04 ($2\frac{1}{100}$) inches.

Q. What would be the pressure per square inch of
the water one mile below the surface of the ocean?

A. It is estimated to be about one ton to the square
inch.

Q. How far below the surface of the ocean does the
wave motion cease to be felt?

A. About 3,500 feet ; and a few feet below the surface
of the sea the water is of the same temperature all over
the world.

Q. How does the friction of water in pipes increase?

A. Friction of water in pipes increases as the square of the velocity; thus, if in one pipe the water is flowing at the rate of 2 feet per second and in another there is water flowing at the rate of 3 feet per second, the friction in the latter will be more than double that in the former: for the square of 2 is 4, and the square of 3 is 9.

Q. What are the four most notable temperatures for pure water?

A. First: freezing point at the sea level, 32° Fahr.; weight per cubic foot, 62.418 lbs. Second: point of maximum density, 39.1° Fahr.; weight per cubic foot 62.425 lbs. Third: British standard for specific gravity. 62° Fahr.; weight per cubic foot, 62 355 lbs. Fourth: boiling point at sea level, 212° Fahr.; weight per cubic foot, 59.76 lbs.

Q. What is the specific gravity of the average sea water?

A. About 1.028 ($1\frac{28}{1000}$). fresh water at the sea level with a mean temperature being taken as a standard or one.

Q. What is the boiling point of the average sea water?

A. About 213.2° ($213\frac{2}{10}$) Fahr.; and the weight of a cubic foot at a temperature of 62° Fahr. would be 64 lbs.

Q. What two ways does water hold substances of a foreign nature?

A. Water holds foreign matter both in solution as well as in suspension.

Q. Can this foreign matter be filtered from the water so as to thoroughly purify it?

A. The particles that are held in suspension can be filtered out; but the matter held in solution cannot be separated except by evaporation, though a great portion can be separated by freezing the water to ice.

Q. What does Rossett consider the most probable temperature of water at its maximum density?

A. About 4.107° ($4\frac{107}{1000}$) Cent. or 39.35° ($39\frac{35}{100}$) Fahr.

Q. Where does the water come from that supplies our principal rainfall (Atlantic States)?

A. From the Gulf and the Gulf Stream : the surface of this water being so much warmer than other waters that evaporation takes place quicker there.

Q. What is the average rainfall of Boston and vicinity?

A. About 48 inches per year; and 28 inches the minimum rainfall.

Q. What is the yearly evaporation from a surface of water in the vicinity of Boston?

A. From experiments made at Chestnut Hill reservoir by Mr. Desmond Fitz Gerald, it was found that the yearly evaporation averages about 35 inches.

Q. What has the greatest specific heat, or heat-absorbing capacity?

A. Water has the greatest specific heat, or heat-absorbing capacity (bromine and hydrogen excepted), and is the unit of comparison employed for all measurements of the capacities for heat of all substances whatever.

Q. What is the least possible pitch that will give motion to water?

A. It is said that an inclination or pitch of one inch in 15 miles would be sufficient to give motion to water if it were possible to construct such a plane.

Q. What velocity in the average river would a pitch of 3 inches to the mile give?

A. About 3 miles per hour; and a pitch of 3 feet to the mile would produce a torrent.

Q. About what is the fall of the Hudson and Mississippi rivers?

A. According to the United States coast survey reports, the fall of the Hudson River from Albany to its mouth is only about 5 feet; and the Mississippi River from its source at Lake Glazier to the Gulf of Mexico is 1,582 feet; from Cairo to the Gulf it is only 322 feet, the greater part of the fall being in its upper end.

Q. What is the coldest body of fresh water known?

A. Lake Superior is the coldest as well as the largest body of fresh water on the globe.

Q. What do we know about water?

A. Water is a liquid; specific gravity, one or unity. It is formed by the chemical union of the two gases, hydrogen and oxygen, in the proportion of two volumes of hydrogen to one of oxygen; or, by weight, of one part of hydrogen to eight parts of oxygen. It exists, in nature, in the three states: solid, as ice or snow; liquid, as water; gaseous, as fog or vapor. Water freezes at 32° Fahr., and boils at 212° Fahr. at the sea level. Its greatest density is at about 39.2° Fahr.; from this point it expands both ways. It is the only single substance known that does not always expand with heat; in freezing it is estimated that it expands from $\frac{1}{11}$ to $\frac{1}{12}$ in volume. It is the most powerful solvent known, as it dissolves minerals, vegetables and gases. On account of its solvent power, water is never obtained pure except when freshly distilled. The presence of salt in water raises the temperature of the boiling point, and lowers that of the freezing point.

SECTION III.

FUEL.

Q. How is heat derived?

A. Heat is derived (artificially) in the most common form by the combining of the two gases, oxygen and hydrogen, with carbon, which is a solid.

Q. What is it in the fuel that gives us heat?

A. The two elementary bodies to which we owe the eating power of all our fuels are carbon and hydrogen.

Q. Which gives off the most heat?

A. According to bulk carbon gives off the most heat, but according to weight hydrogen gives the most heat; for one pound of carbon, it is said, will heat 14,220 lbs. of water 1° Fahr., while one pound of hydrogen will heat 52,155 lbs. of water 1° Fahr.

Q. Which will ignite first, carbon or hydrogen?

A. Hydrogen will ignite first at a temperature of about 300° Fahr., while carbon requires about 1.000° of heat to ignite it (a low red lustre), and even then burns very slowly.

Q. What is the lowest possible temperature of a furnace when combustion is going on?

A. Never less than 1,000° Fahr.

Q. How much heat will the average match generate?

A. Probably about 700° Fahr.

Q. What is the standard instrument for measuring heat?

A. The standard instrument for measuring heat in Britain and America is the Fahrenheit thermometer. Its boiling point is 212°, and its freezing point is 32°: this makes just 180° between freezing and boiling points. For scientific purposes the French instrument is used nearly always. (This is the Celsicus, or more commonly called the Centigrade.) The freezing point on this instrument is 0°, and the boiling point is 100°, making just 100° between freezing and boiling points.

Q. How would you change a Cent. temperature to a Fahr. temperature?

A. To change a Cent. temperature to a Fahr., simply multiply the Cent. temperature by $\frac{9}{5}$ and add 32. Reverse the rule to change back; thus, if we wish to change 100° Cent. to Fahr., we proceed thus: $100 \times \frac{9}{5} = 180 + 32 = 212$.

Q. What is the substance used in an ordinary thermometer?

A. The substance used in an ordinary thermometer is mercury. It is the heaviest of all liquids. It will expand $\frac{1}{9990}$ of its volume for each Fahr. degree of heat applied, and it will freeze at a temperature of 37.8° ($37\frac{8}{10}$) below zero; for colder temperatures alcohol is used, it being the hardest substance known to freeze (in its pure state).

Q. What will be the pressure at the base of a column of mercury 30 inches high?

A. Just 15 lbs. per square inch.

Q. What temperature does mercury boil at?

A. Mercury boils at a temperature of 662° Fahr. For higher temperatures the use of the pyrometer is substituted. In this instrument the heat is measured by the expansion of metals, and it is said that it will register correctly up to 7,000° Fahr.

Q. Do all substances expand with heat?

A. About all single substances do.

Q. What would be the hardest substance of all to melt?

A. Carbon would be the hardest substance of all to melt.

Q. What is meant by a unit of heat?

A. It is an amount of heat required to raise one pound of water one degree Fahr.

Q. What is meant by combustion?

A. Combustion is the term applied to the process of burning, which usually consists of the oxygen of the air uniting with the constituents of the combustible substances. Thus the combustion of fuel is due to the oxygen of the air passing into a state of chemical union with the carbon and hydrogen of the fuel, forming carbonic acid (CO_2) and water vapor (H_2O).

Q. Does such chemical combination always generate heat?

A. Yes, always.

Q. Does the oxygen of the air have as much to do with combustion as carbon and hydrogen?

A. Yes; it is usually styled the supporter of combustion.

Q. How much oxygen is necessary for the combustion of one pound of coal?

A. 156 cubic feet of air must pass through the grate for every pound of coal consumed, and about one-fifth part of the air is oxygen.

Q. What would be the effect of excess of air on combustion?

A. The effect of excess of air on combustion, either with carbon or hydrogen as fuel, has been proven to have no effect on the quantity of heat produced where combustion was perfect; but the intensity of temperature will be diminished in any one place, for the fire covers a larger area according as the draft is increased.

Q. What temperature does the phosphorus of a match inflame at?

A. At 150° Fahr.; so the mere friction of scratching the match generates heat enough to ignite it.

Q. What temperature does the sulphur of the match ignite at?

A. At about 500° Fahr.

Q. What temperature does wood ignite at?

A. At about 800° Fahr.

Q. What temperature does coal ignite at?

A. Coal will ignite perfectly at 1,000° Fahr.

Q. What temperature will kerosene oil freeze at?

A. At from 10° to 12° below zero, Fahr.

Q. What is the weight of one gallon of the crude petroleum?

A. About 7.3 ($7\frac{3}{10}$) lbs.

Q. What is meant by latent heat?

A. Latent or hidden heat is the heat that disappears in changing ice to water, or water to steam. This heat is not perceptible on the thermometer, for the thermometer only registers the sensible heat of bodies.

Q. What substance will render latent the greatest amount of heat?

A. Water, it is said.

Q. What is the boiling point of fresh water?

A. In a complete vacuum, at a temperature of 72° Fahr., under the ordinary atmospheric pressure at the sea level at a temperature of 212° Fahr., and, roughly speaking, for every 520 feet in height we ascend the boiling point will be reduced one degree. Water will boil only when it reaches a temperature of 234° Fahr. under a pressure of 15 lbs. per square inch.

Q. How would the boiling point of other liquids be affected by a vacuum?

A. In a complete vacuum liquids in general boil at a temperature of 140° Fahr. lower than in the open air.

Q. What is the average rate of increased heat of the earth as we penetrate down below its surface?

A. About one degree Fahr. for every 64 feet in depth.

Q. What is the relative value of some of the most common of our fuels?

A.

1 lb. Charcoal, pure,   will raise 78 lbs. water from 32° to 212° Fahr.
1 "   "   from wood, "   75 "   "   "   "   "   "   "
1 " Wood, dried,   "   36 "   "   "   "   "   "   "
1 "   "   undried,   "   27 "   "   "   "   "   "   "
1 " Coal, bituminous,   "   60 "   "   "   "   "   "   "
1 " Turf and Peat,   " 25 to 30 "   "   "   "   "   "   "
1 " Alcohol,   "   67½ "   "   "   "   "   "   "
1 " Olive Oil, Wax, etc., " 90 to 95 "   "   "   "   "   "   "
1 " Ether,   "   80 "   "   "   "   "   "   "
1 " Hydrogen,   "   236½ "   "   "   "   "   "   "

Q. How many units of heat disappear in the conversion of one pound of boiling water into steam?

A. It is estimated that there are about 1,004 Fahr. units of heat that disappear in the conversion of one pound of boiling water into steam, and yet the steam will only have the same amount of sensible heat as the water, for there is so much heat that disappears or becomes latent in changing water to steam.

Q. How many cubic feet in the average ton of coal (2,000 lbs.)?

A. Allow just 32 cubic feet per ton of average coal.

Q. What will be the weight of air required to burn one pound of coal?

A. It takes 32 parts by weight of oxygen to consume 12 parts by weight of carbon. This is the proper proportion of each to produce carbonic acid gas, which we wish to do in order to generate the greatest amount of heat. It takes 4.35 lbs. of air to supply one pound of oxygen; therefore, it will take 11.5 lbs. of air to provide the gas essential to the economical combustion of one pound of coal.

Q How does wetting coal add to its fuel value?

A. Heat resolves the moisture into steam, and finally into carbonic oxide and hydrogen. If the draught of air supplied to the fire is sufficient, both these gases will burn. The injection of steam will accomplish the same purpose.

2

Q. What advantages arise from the use of petroleum as fuel, leaving the cost of the fuel out of the question?

A. Fewer men at the fires, no smoke or soot, no ashes, fires can be kept more regular and are under better control, steam is more uniform.

. *Q.* Setting aside the cheapness of petroleum, what advantages does it possess?

A. The having absolutely perfect control over the fire, as well as its great cleanliness.

Q. What is the theoretical evaporation due to the combustion of one pound of coal?

A. Fifteen pounds of water from a temperature of 212° Fahr.

Q. What is the theoretical evaporation due to the combustion of one pound of petroleum?

A. About twenty-two pounds of water, varying somewhat with the kind of oil.

Q. What is about the best practical value of one pound of coal ?

A. From ten to eleven pounds of water.

Q. What is the best practical value of one pound of petroleum?

A. About eighteen pounds of water.

Q. What proportion of the space occupied by one ton of coal would be required to stow an amount of petroleum of equal total evaporative value?

A. About three-eighths ($\frac{3}{8}$) of the space occupied by the ton of coal.

Q. One ton of the average coal is equal to about how much wood for steaming purposes?

A. About two cords of the average wood.

Q. How many primary divisions can we divide coal into?

A. Two primary divisions, namely : anthracite, or hard coal, which does not flame when kindled ; and bituminous, or soft coal, which does. The reason is, the soft coal contains so much more hydrogen and ignites at so low a temperature, that it flames the instant it touches a hot fire.

Q. How much carbon does anthracite coal contain?

A. Anthracite sometimes contains as much as 94 per cent of carbon; and as this element decreases in amount, it graduates into a bituminous coal. The term " anthracite " is never applied to coal containing less than 80 per cent of carbon.

Q. How many varieties does bituminous coal contain?

A. Bituminous coal includes almost an endless number of varieties, one of the costliest being cannel coal. Cannel coal contains more gas than any other; in fact, it contains from 8,000 to 15,000 cubic feet per ton.

Q. What does a ton of average coal contain?

A. It is said that a ton of coal contains, besides the gas, 1,500 lbs. of coke, 20 gallons of ammonia water and 140 lbs. of coal tar.

Q. In using coal, what portion of it burns on the grate?

A. Fifty to 67 per cent of carbon (which is the solid part) burns on the grate, while 20 to 32 per cent of carbon and hydrogen (the gas) has to burn in the open space above and back of the fuel, or escape unconsumed.

Q. How much gas will be required to generate as much heat as the average coal per ton?

A. The heating power of one ton of the average coal is equal to 40,000 cubic feet of gas.

Q. How much water will a pound of coal evaporate?

A. It takes about one pound of the average coal to evaporate one gallon of fresh water in the average boiler.

Q. What is the maximum consumption of coal per square foot in steam boilers?

A. The consumption of coal for steam boilers cannot exceed (with natural draught) 12 lbs. per hour for each square foot of grate surface.

Q. How do we measure the value of fuels?

A. The value of any fuel is measured by the number of heat units which its combustion will generate. The fuel used in generating steam is composed of carbon and hydrogen and ash, with sometimes small quantities of other substances not materially affecting its value. " Combustible " is that portion which will burn; the ash or residue varying from 2 to 36 per cent in different fuels.

Q. How many sources of waste are there in fuels?

A. There are two sources of waste in fuels burned under steam boilers. The gases going to waste to the chimney carry off on an average 31 per cent of fuel, and in the most economical boilers this cannot be reduced below 12 per cent; then the feed water has to be heated from the normal temperature to that of the steam before evaporation can commence, and this generally at the expense of the fuel which should be utilized in making steam.

Q. What can we say in regard to petroleum as fuel?

A. Crude petroleum is, without doubt, the coming fuel on locomotives and ocean steamers; for less than one-half the room formerly used for coal will be required to store the oil, while the weight will not be quite one-half. Three barrels of oil of 42 gallons each slightly exceeds the heating capacity of a ton of average coal. The oil weighs 913½ lbs. and may be purchased at a saving of about 50 per cent, it is said.

Q. Is there any fuel that would be cleaner than oil to handle?

A. Gas is the only fuel that would be cleaner.

Q. How much air will be required to burn one pound of coke?

A. One hundred and fifty-six cubic feet of air is required to pass through the grates to burn one pound of average coal; and coke, it is said, requires one-third more.

Q. Which is the best, cold or hot air for a boiler furnace?

A. The coldest air, if thoroughly mixed on its entrance with the fuel or gases, will never cool, but will always sustain or increase the heat of the furnace. Air in bulk only can do any harm; and this is objectionable from the obstruction which it forms to combustion, as well as from its abstraction of heat from the furnace. Thus, the air, divided into thin streams, should be taken from the outside of the boiler directly into the furnace through very small holes or openings.

SECTION IV.

STEAM.

Q. What is steam?

A. Steam is an elastic fluid generated by the action of heat upon water.

Q. What is steam composed of ?

A. The vapor arising from water at or above its boiling-point, called "steam," is a chemical compound consisting of eight parts by weight of oxygen and one part of hydrogen. Steam proper is perfectly transparent and colorless, dry, and only moist when condensed, wholly invisible, and when apparent, only so by reason of partial condensation.

Q. What is the weight of steam?

A. 26 36 $(26\frac{36}{100})$ cubic feet of steam at the atmospheric pressure weighs just one pound (avoirdupois) ; or 5 cubic feet of steam at 75 lbs. pressure to the square inch (gauge pressure) weighs about one pound.

Q. What is meant by low pressure steam?

A. Low pressure steam is steam not exceeding 15 lbs. per square inch.

Q. What is meant by superheated steam?

A. Superheated steam is steam which has a greater temperature than that due to its pressure; that is, heat is applied to the steam pipes or vessel containing steam after it has left the water from which it was generated.

Q. Does the pressure of steam increase as the temperature?

A. The pressure of steam increases at a far higher rate than the temperature; doubling the temperature increases the pressure nearly twenty-three times.

Q. How do we find the latent heat of steam?

A. The latent heat of steam is generally found by subtracting its sensible heat from 1,202; at 45 pounds it will be exactly right, for at that pressure the total heat will be just 1,202° Fahr.

Q. Does steam or vapor rise from water at all temperatures?

A. It does.

Q. Is there any difference between steam and vapor?

A. The chemical analysis shows no difference, but the different authorities generally consider them different; for, usually speaking, steam is formed artificially, while vapor is formed naturally; and another difference is the temperature — anything less than 212° Fahr. is styled vapor, and over that it would be called steam.

Q. Is there any rule for the amount of steam pipe required to heat a building with steam?

A. The "Master Steam Fitter" gives three rules, as follows, but I consider the third one the most practicable :

1st. One square foot of steam pipe for every 6 square feet of glass in the windows.

2d. One square foot of steam pipe for every 100 square feet of wall and ceiling.

3d. One square foot of steam pipe for every 80 cubic feet of space to be heated.

In heating a room with hot water a larger surface to radiate from is needed, on account of its lower temperature.

Q. Suppose we have seven stores to heat with steam, whose dimensions are each 100 feet long. 30 feet wide and 15 feet high. How many square feet of steam will be required to heat them, and how many running feet of 1-inch pipe will it take. the 1-inch pipe being about 4 inches in circumference ?

A. Figure thus :

$$100 \text{ ft. length of each store.}$$
$$15 \text{ ft. height } `` \qquad ``$$

$$\begin{array}{l} 500 \\ 100 \end{array}$$

1500
        30 ft. width of each store.

45000 cubic ft. of space in one store.

No. cubic ft.   7 number of stores.
that 1 sq. ft.   ———
of steam pipe } 80)315000 cubic ft. of space in the 7 stores.
will heat.

        3937 + No. sq. ft st'm pipe req'd to heat 7 stores.
        3 ft. the length of 1-inch pipe required to get 1
        ——— sq. ft. h't'g surface from, as proven below.
        11811 ft. length of 1-inch pipe to heat the 7 stores.
Circumf. of a 1-inch pipe 4)144 sq. inches to 1 sq. ft.

        ———
        36 = 3 ft. length of 1-inch pipe to get
                1 sq. ft. of heating surface in.

Q. How does steam absorb what is called "latent" heat?

A. When steam is generated in a boiler the water is heated until it arrives at the boiling-point (212° Fahr.); and if the vessel is an open one, the temperature cannot be raised any higher. But if we wish to convert all the water into steam, we will have to add a great deal more heat, and this is the heat that disappears, or latent heat as it is called; for the steam rising from boiling water in an open vessel is of the same temperature as the water, sensibly, but still we know it contains a great deal more heat than the water (sensible and latent heat combined).

Q. How much lighter is steam at 212° Fahr. than air?

A. Steam at 212° Fahr. is not quite one-half as heavy as air, the specific gravity of air being 1.0000 and that of steam at 212° Fahr. but .4883 ($\frac{4883}{10000}$).

Q. In running an engine, which is the most economical, high or low pressure steam?

A. High pressure steam is the most economical, because we have a greater expansion force to profit by.

Q. In heating a building, which is the most economical, high or low pressure steam?

A. Low pressure steam; for the relative volume of steam decreases faster than the temperature increases, as the pressure rises.

Q. What is meant by the relative volume of steam?

A. It is the proportional amount of steam that a certain amount of water will produce.

Q. Give a table of the relative volume of steam at different pressures; also the temperatures.

Steam.	Temperature in Fahr.	Relative volume.
	212°	1700
10 lbs. per sq. inch.	240°	1040
15 " " "	250°	903
20 " " "	260°	765
25 " " "	267°	677
30 " " "	274°	608
35 " " "	281°	552
40 " " "	287°	506
45 " " "	293°	467
50 " " "	298°	434
55 " " "	303°	406
60 " " "	308°	381
65 " " "	312°	359
70 " " "	316°	340
75 " " "	320°	323
80 " " "	324°	307
85 " " "	328°	293
90 " " "	332°	281
95 " " "	335°	269
100 " " "	338°	259
105 " " "	341°	249
110 " " "	344°	239
115 " " "	347°	231
120 " " "	350°	223
125 " " "	353°	216
130 " " "	356°	209
135 " " "	358°	203
140 " " "	360°	197
145 " " "	363°	191
150 " " "	365°	186

SECTION V.

BOILERS.

Q. What is a steam boiler?

A. The boiler is the vessel in which steam is generated, to furnish motive power for the steam engine.

Q. What are some of the most common boilers used?

A. The horizontal return tubular boiler, the horizontal return flue boiler, the upright tubular boiler, the locomotive style of a boiler, and the safety or sectional boiler.

Q. Which is the most common boiler used?

A. The horizontal return tubular boiler is the most common one used; and it is the best when we take into consideration the first cost, repairs, economy of fuel, the ease with which it can be cleaned, the length of time it will last, and the ease with which it can be handled.

Q. What is the furnace?

A. It is the space above the grate where the fire lies.

Q. What is the ash-pit?

A. It is the space below the grate.

Q. What is the bridge wall?

A. It is the wall at the back end of the grate in the return tubular or flue boilers.

Q. What is the use of the bridge wall?

A. It keeps the coal from falling off the rear end of the grate. It also forces the flame up to the bottom of the boiler; and, probably the most important of all, when the doors are opened to put on fresh fuel it reduces the amount of cold air that will be drawn out under the boiler back through the tubes or flues and up the chimney.

Q. What is the combustion chamber?

A. It is the space back of the bridge wall.

Q. What is the blow-off pipe?

A. It is a pipe put in at the bottom or lowest part of the boiler for emptying the boiler when necessary, or to open occasionally to blow out the sediment that may accumulate.

Q. What is the man-hole in a boiler?

A. It is an opening for a person to enter the boiler to inspect or repair it.

Q. What are the brackets on a boiler?

A. They are the castings riveted to the sides of a boiler for holding it up on the brick work.

Q. What are the stays in a boiler?

A. The braces for holding the flat places from bulging out.

Q. What are the hand-holes in a boiler?

A. Small openings for cleaning or inspecting the boiler.

Q. What is the safety plug in a boiler?

A. It is a plug of metal that will fuse at a low temperature, placed at the low water line in a boiler.

Q. What is the use of the safety plug?

A. If the water should be allowed to get down as low as the safety plug, it is pretty apt to melt out and put out the fires; and, if it works properly, the boiler could not explode from low water. But, of course, there are several reasons why it could not always be depended upon; and it is not every boiler that has one.

Q. What is meant by the heating surface (H. S.) in a boiler?

A. The H.S. of a boiler is that portion exposed to the fire, and, of course, it must be always covered with water.

Q. How many square feet of H. S. will it take to equal one horse power (H. P.)?

A. In our best boilers, which are the horizontal return tubular, we allow 15 square feet of H. S. per H. P.

Q. What will be the H. P. of a boiler of the following dimensions: 18 feet long, 5 feet diameter, with 74 tubes that are each 3 inches in diameter? (This boiler is of the

horizontal return tubular pattern, and we will suppose
the H. S. of the shell to include one-half of the circumference of the boiler.)

A.   Proceed thus :

```
        3.1416 decimal number to get circumf. from diam.
           5 ft. diameter of the boiler.
        ─────
    2)15.7080 ft. circumference of the boiler.
       ─────
       7.854 ft. ½ circumference of the boiler.
         18 ft. length of the boiler.
       ─────
       62832
       7854
       ─────
     141.372  sq. ft. of H. S. in the shel' of the boiler.
    1046.1528    "    "    "      "   74 tubes of the boiler.
```

Sq. ft. of ⎫
H. S. per ⎬ 15)1187.5248   "   "   "   "   whole boiler.
H. P.   ⎭

```
       79.1683 H. P. of a boiler of the above dimensions,
                 but it is so near 80 that we would call it
                 an 80-H. P. boiler.
        3.1416 decimal number to get circumf. from diam.
           3 inches diameter of each tube.
```

Inches per ft. 12)9.4248 inches circumference of each tube.

```
        .7854 ft.           "              "          "
          18 ft. length of each tube.
       ─────
       62832
       7854
       ─────
      14.1372 sq. ft. H. S. in 1 tube.
           74 number of tubes in the boiler.
       ─────
       565488
       989604
       ─────
     1046.1528 sq. ft. of H. S. in the 74 tubes.
```

Q.   How do we find the safe working pressure (S. W.
P.) of a boiler?

A.   The rule that the United States inspectors and the
insurance inspectors use is : Take the tensile strength
of the boiler plates and divide it by 6, multiply this quotient by the thickness of the plate in fractional parts of an
inch, then take this product and divide it by the radius of
the boiler in inches ; this quotient will be the S. W. P. of
a single-riveted boiler, if reasonably new. If the boiler is
double-riveted, 20 per ct. can be added to the above result.

Q. What is meant by the tensile strength?

A. The strength required to tear asunder a piece of metal whose cross sectional area will be equal to one square inch. Take, for example, a piece of metal one-half of an inch thick and 2 inches wide, or a piece one-fourth of an inch thick and 4 inches wide, will have a cross sectional area of 1 square inch.

Q. What is meant by the radius of a boiler?

A. One-half of the diameter is the radius.

Q. Get the S. W. P. of a boiler whose tensile strength is 40,000 lbs., thickness of plate one-fourth of an inch, and diameter of shell 54 inches.

A. Proceed thus:

Constant 6)40000 lbs. tensile strength of the iron.

$$\frac{6666}{.25}= \text{¼ of an inch, thickness of the iron.}$$

$$\frac{33330}{13332}$$

Radius of
boiler } 27)1666.50(61.7=61$\frac{7}{16}$ lbs., the S. W. P. of the boiler if
in inches, }   162   single-riveted and in good
                             condition.

$$\frac{46}{27}$$

$$\frac{195}{189}$$

6

Q. If the tensile strength of a boiler was not known, how would we get the S. W. P.?

A. For an iron boiler we would take 40,000 lbs. as a tensile strength, because that is the lowest strength of iron used in the construction of boilers (50,000 lbs. is the highest). For a steel boiler we would use 50,000 lbs. of tensile strength, because that is the lowest tensile strength of steel used in the construction of boilers (65,000 lbs. is about the highest).

Q. What proportion of the strength of the whole boiler plate is the strength of a single-riveted joint?

A.   56 per cent of the strength of the whole plate.

Q.   What is the strength of a double-riveted joint?

A.   70 per cent as strong as the whole plate.

Q.   What is the strength of a triple-riveted joint?

A.   82 per cent as strong as the whole plate.

Q.   How should an ordinary return tubular boiler be set?

A.   On a good, firm foundation, so that it will not settle in any way; and it is a good plan to set the rear end a little low, say from one-half to three-quarters of an inch for an ordinary length boiler (14 to 18 feet).

Q.   What is the best length for grate bars in an ordinary horizontal return tubular or flue boiler?

A.   From 4 to 5 ft. in length gives about the best results.

Q.   Can we reckon the H. P. of a boiler by the number of square feet of grate surface (Gt. S.) it contains?

A.   Yes; for each square foot of Gt. S. in our modern return tubular boilers will be equal to 3 H. P.; and each square foot of Gt. S. in the locomotive style of a boiler, or the upright tubular boiler, will be equal to 4 H. P. These rules apply to boilers where natural draught alone is depended upon for the combustion of the fuel.

Q.   How do we find out how many square feet of Gt. S. there should be under a boiler?

A.   In our best modern boilers (horizontal return tubular) the diameter of the boiler in feet multiplied by itself will equal the Gt. S. in square feet; or, divide the H. P. of the boiler by 3, the quotient will equal the Gt. S. And still another way is, to divide the H. S. by 45; the quotient will equal the Gt. S. in square feet.

Q.   How is the quickest way to find out the number of square feet of H. S. in a boiler?

A.   Multiply the number of square feet of Gt. S. by 45; the product will equal the H. S. nearly.

Q.   How much area of space do we leave over the bridge wall under a return tubular or flue boiler?

A.   18 square inches of space per H. P. of the boiler.

Q. What should be the total area of the flues or tubes in a boiler?

A. The total area of the flues or tubes in a boiler is from one-fifth to one-seventh the area of the grate.

Q. What should be the area of a chimney for a boiler?

A. From one-seventh to one-tenth the area of the grate.

Q. In putting a fire under a boiler where everything is cold, is there any danger of heating up too quickly?

A. There is, for there will be an unequal expansion; and we do not want to bring the change about too quickly, for there would be more danger of injuring the boiler in so doing.

Q. If we were going to clean a boiler that was set in brick-work, what is the most important thing to look after?

A. Do not blow the water off till the brick-work has had time to cool off some. For an ordinary boiler, let it stand with the dampers open for at least 4 or 5 hours after the fire has been all drawn out before letting the water off.

Q. Why should leaks be stopped as soon as possible?

A. The principal reason is on account of the loss of steam that takes place, then the wearing away of the parts that come in contact with the leak, and the sound of escaping steam which is very disagreeable to the ear.

Q. How near the under side of the boiler should the grate bars be, to give the best results?

A. Grate bars should be about 24 inches below the lowest part of the boiler for burning anthracite coal, or any fuel that does not contain a large proportion of volatile matter (smoky substance). When soft coal is used, from 27 to 30 inches is found to give the best results.

Q. What is the most important fixture on a steam boiler?

A. The safety-valve; for, when it is in good working order (if it is of sufficient size), there will be no possibility of getting an over-pressure of steam.

Q. What is the rule for getting the size of a safety-valve?

A. One-fifth of the H. P. of the boiler will equal the area of the safety-valve in square inches, or the Gt. S. multiplied by .5 to .8 ($\frac{5}{10}$ to $\frac{8}{10}$). Use the larger fraction for the self-contained fire-box boilers, as the locomotive style, and the upright boilers.

Q. What is the greatest danger that can happen to a safety-valve?

A. Getting corroded, or stuck down, or overloaded.

Q. Will a safety-valve close at the same pressure that it opens?

A. No; for there is a little more surface exposed to the steam when it is open; usually there is about 3 lbs. difference between opening and closing points.

Q. How do we keep a safety-valve in good working order?

A. A safety-valve should be tried at least once every day in the following manner: Let the steam run up to the blowing point and see if the valve opens promptly. If it does, it shows that the valve is all right, and all we have to do is to speed up the pump, or close the dampers to put the steam down so there will not be too much steam lost. If the valve does not start itself, then it will have to be raised by hand carefully.

Q. What is the rule for the size of a safety-valve that the Board of Trade uses?

A. For return tubular boilers the Board of Trade allows one-half of a square inch area of safety-valve for each square foot of grate surface.

Q. What rule do the insurance inspectors of this country use for size of safety-valves?

A. They give for a pop, or spring-valve, 1 square inch area for every 3 square feet of grate surface; and for a common valve, 1 square inch area for every 2 square feet of grate surface.

Q. How do we get the number of square inches in a safety-valve?

A. Multiply the diameter of the valve in inches by itself, then multiply the product thus obtained by .7854 ($\frac{7854}{10000}$); this last product will be the square inches that the valve contains.

Q. How do we get the total pressure on a safety-valve?

A. Multiply the weight on the lever in pounds by the distance in inches that it sets from the stud or fulcrum, and divide the product by the distance in inches from the center of the stud or fulcrum to the center of the valve stem, in line with the lever; this quotient will be the total pressure on the valve. Now divide the total pressure on the valve by the number of square inches area in the valve, and the quotient will be the pressure per square inch on the boiler, or " gauge pressure," as it is called.

Q. How do we get the length of the lever of a safety-valve?

A. Take the total pressure on the valve and multiply it by the distance in inches from center of stud to center of valve stem; now divide this product by the weight of the ball in pounds; the quotient thus obtained will be the distance in inches from the center of the weight on lever to the center of the stud or fulcrum, or " the length of the lever," as it is called.

Q. How do we find the weight of the ball on a safety-valve?

A. Take the total pressure on the valve and multiply it by the distance in inches from the center of the stud or fulcrum to the center of the valve stem in line with the lever; now divide this product by the length of the

3

lever in inches (from center of stud or fulcrum to center of weight on lever); the quotient thus obtained will be the weight of the ball in pounds.

Q. How are spring, or pop, safety-valves adjusted?

A. Spring-valves, or what are usually called "pop-valves," are adjusted by a test gauge, and are usually locked up, so the engineer cannot tamper with them. When they are not locked up I consider them a dangerous valve.

Q. What will be the total pressure on a safety valve of the following dimensions: Valve, 3 inches in diameter; weight or ball, 105 lbs.; distance from center of ball on lever to center of stud, 20 inches, and distance from center of stud or fulcrum to center of valve stem (in line of lever), 3 inches?

A. Proceed thus:

```
              3 inches, diameter of valve.
              3 inches    "      "    "
              ―
              9 circular inches, area of valve.
            .7854
            ―――――
         7.0686 square inches, area of valve.
            105 lbs., weight of ball.
             20 inches, length of lever.
Distance from )     ――
stud to valve, ) 3)2100
                 ―――――
                 700 lbs., total pressure on the valve.
Sq. in., area )
in the valve, ) 7.0686)700.0000(99. lbs. pressure per sq. in. on boiler
                636174                  when valve opens.
                ―――――
                638260
                636174
                ―――――
                 2086
```

Q. What will be the length of lever of the above valve?

A. Proceed thus:

```
              700 lbs., total pressure on the valve.
                3 in. distance from stud to valve stem.
                ――
Weight of the ball,105)2100(20 in., length of lever from stud to weight.
                        210
                        ―――
                          0
```

Q. What will be the weight of the ball of the above valve?

A. Proceed thus :

700 lbs., total pressure on the valve.
3 in. dist'ce from center of stud to center of valve stem.

Length)
of   } 20)2100(105 lbs., the weight of the ball.
lever,)   20

100
100

Q. How do we prove a safety-valve?

A. We can prove it by means of the above figures, or by means of the steam gauge; for, of course, the steam gauge and safety-valve must always prove each other.

Q. What must we observe in relation to the steam gauge?

A. The steam gauge must stand at zero when the pressure is off, and it should show the same pressure as the safety-valve is set to blow at when the safety-valve blows.

Q. What means do we have of knowing the height of the water in a boiler?

A. The most common contrivances used are the water glass and gauge cocks. The latter is considered the most reliable: but I think any intelligent person will choose the water glass.

Q. How can we tell if the glass is all clear and right?

A. By the motion of water in the glass, by the blow-off, or by means of the feed pump.

Q. Which end of the water glass is the most liable to get clogged up?

A. The lower end of the glass is the most liable to get clogged up with mud. It is the same with the lower gauge cocks.

Q. If the water glass breaks, what should we do?

A. Close the valve in the lower end of the glass first; then, when that is done, close the upper one. In this way a person is not very apt to get burned. '

Q. How much space will one horse-power of a boiler heat?

A. One horse-power of a boiler is sufficient to heat 40,000 cubic feet of space.

Q. How do we get the pressure on stay-bolts in a boiler?

A. To get the pressure on stay-bolts in a boiler where they are set in squares (as in the fire-box of a locomotive or portable boiler), simply multiply the distance between stays, center to center, by itself; this equals the number of square inches area that each stay has to support, minus the area of one stay. Now multiply this by the boiler pressure; this will equal the pressure on each stay.

Q. Suppose in a locomotive boiler the stays in the fire-box are 1 inch in diameter and 5 inches apart (center to center), with a boiler pressure of 130 lbs. to the square inch; what will be the pressure on each stay?

A. Proceed thus :

5 in. apart, center to center.
5 "   "   "   "   "
——
25 sq. in. in area, area of 1 stay.
.7854 sq. in., area of 1 stay.
——
24.2146 sq. in., area 1 stay supports.
   130 lbs. pressure per sq. in.
——
 7264380
 242146
——
3147.8980 lbs., pressure 1 stay has to support.
   1 in., diam. of stays.
   1 "   "   "   "
——
   1
.7854 times square of diameter in inches.
——
.7854 sq. in., area of stay.

Q. How much pressure per square inch will these stays stand with safety?

A. The material used in the construction of these stays will safely stand 5,000 lbs. per square inch of cross sectional area.

Q. What is the consumption of water per H. P. of a boiler?

A. Each nominal H. P. of a boiler requires one-half of a cubic foot (about 30 lbs.) of water per hour.

Q. What is the evaporation of water per pound of coal in boilers?

A. From 7 to 10 lbs. of water per pound of average coal. Take the average boiler, and the evaporation will be a gallon (8½ lbs.) of water to a pound of coal.

Q. What is the best substance to put in a boiler to keep it clean?

A. Soda is one of the best as well as one of the cheapest substances for keeping a boiler clean, and it is recommended by the inspectors.

Q. How is the soda used in a boiler?

A. If a boiler is very dirty, dissolve and pump in from one to three pounds per day; blow off some water every morning before the fire has been started; let the boiler down and clean often. If a boiler is clean and the water used good, a couple of pounds per week will be sufficient.

Q. Where should the blow-off be attached to a boiler?

A. All boilers should have both a surface blow-off, and one at the bottom or lowest part of the boiler; the surface blow-off to be used while running, for the water is then in motion; the lower one to be used in the morning, or at such time as the water has been at rest long enough for the sediment to settle to the bottom.

Q. How much more heat will be required to heat through ordinary boiler scale than iron of the same thickness?

A. The heat-conducting power of ordinary boiler scale compared with that of iron is as 37 to 1; or, taking it in another way, scale $\frac{1}{16}$ of an inch thick over the H. S. of a boiler will require an expenditure of 15 per cent more fuel than if the same boiler was clean.

Q. Give proportions of riveted joints or seams in a boiler.

A.

	Double-riveted.	Single-riveted.
Thickness of plate,	1.0	1.0
Diameter of rivet,	1.7	1.7
Breadth of lap,	8.3	5.4
Pitch of rivet in line,	7.1	4.6
Distance apart of pitch lines,	2.8	
Distance from edge of plate,	2.7	

Q. What gives us draught in a chimney?

A. Other things being equal, the action of a chimney depends upon its perpendicular height and the difference in temperature within and without the chimney.

Q. How do we get the best results from a chimney?

A. To get the best results from a chimney, let every particle of air that enters it pass through the grates.

Q. What kind of a chimney gives the best draught?

A. Brick chimneys are better than iron, for the upward current of gases does not get cooled off so much; and round chimneys are better than square ones, for the gases ascend in a spiral motion.

Q. Would there be any harm if water came in contact with the exterior of a boiler?

A. Yes; look out that there are no leaks in the boiler or roof which might cause the exterior of the boiler to get wet; for corrosion is sure to take place in a short time, if such places are not looked after.

Q. Give good modern proportions of a 96-H.P. boiler of the return tubular pattern.

A. 18 feet, length of the boiler.

6 feet, diameter of the boiler.

1.440 square feet of H. S. in the whole boiler.

1.260   "   "   "   " tubes of the boiler.

180   "   "   "   " shell   "   "

90 tubes in the boiler, each 3 inches in diameter.

32 square feet of Gt. S. in the grate.

12½ " area over the bridge wall under boiler.

4 " " of the chimney.

3S0 lbs. of coal, maximum hourly consumption.

410 gallons of water " " "

80 lbs. of steam, blowing point of safety-valve.

75 " " maximum pressure run.

90 " " S.W.P. of the boiler (single-riveted).

46,900 lbs., the tensile strength of the iron.

$\frac{7}{61}$ of an inch, the thickness of the iron.

16 inches, area of the safety-valve.

4½ inches, diameter of the safety-valve.

Q. What about water-glasses and gauge cocks?

A. Water-glasses and gauge-cocks should be kept clean, both within and without. They should be frequently blown out to make sure the passages are all clear.

Q. What about a water column?

A. When a water column or combination is used it should be frequently blown out to make sure that it does not get stopped up with mud or scale. It is the water passages where trouble is liable to arise.

Q. What is the fusible plug used for in a steam boiler?

A. The fusible or safety plug is an extra precaution in a boiler, as a guard against low water.

Q. What is the composition and melting points of the ordinary safety plug?

A.

Tin,	6 parts.	Lead,	1 part.			Melts at 381° Fahr.	
"	5 "	"	1 "			" 378°	"
"	4 "	"	1 "			" 365°	"
"	3 "	"	1 "			" 356°	"
"	2 "	"	1 "			" 340°	"
"	1½ "	"	1 "			" 334°	"
"	4 "	"	4 "	Bismuth,	1 part.	" 320°	"
"	3 "	"	3 "	"	1 "	" 310°	"
"	2 "	"	2 "	"	1 "	" 292°	"
"	1 "	"	1 "	"	1 "	" 254°	"
"	3 "	"	5 "	"	8 "	" 212°	"
"	19 "	"	31 "	"	50 "	" 212°	"
"	1 "	"	1 "	"	2 "	" 201°	"
"	2 "	"	3 "	"	5 "	" 199°	"
Zinc,	33.3 parts.	"	33.3 parts.	"	33.4 parts.	" 200°	"

Q. If we were going to select a composition for a fusible plug in a boiler of which the highest steam pressure to be run is 60 lbs., how should we choose one?

A. By referring to our table of temperatures of steam, we find that 60 lbs. of steam has a temperature of 308° Fahr.; and on looking at our table of alloys for fusible plugs, we find that a composition of 3 parts of tin, 3 parts of lead and 1 part of bismuth will melt at a temperature of 310° Fahr. So this will be the one we will select.

Q. How much H. S. should a locomotive boiler have?

A. A locomotive boiler should have five times as many square feet of H. S. as there are square inches area in one of the pistons.

Q. Where should the safety-plug be placed in a locomotive boiler?

A. In the center of the crown sheet, or top of the fire-box; for that is the highest portion of the boiler exposed to the fire.

Q. Where should it be placed in a return tubular boiler?

A. In the rear head of the boiler, just above the top of the upper row of tubes; for in this class of a boiler the top of the upper row of tubes is the highest portion of the boiler exposed to the fire: consequently, it would be the first part bared to the fire if the water was allowed to get low.

Q. Is there any precaution to be taken in regard to the safety-plug when cleaning a boiler?

A. The safety or fusible plug must be examined when cleaning a boiler out, and be carefully scraped on both sides to better facilitate its working, in case the water were to get low in the boiler.

Q. How often should safety-plugs be removed in a boiler?

A. It is a good plan to remove them every two or three years and have them replaced by new ones, as the new ones are a little more reliable than the old ones.

Q. How do we prepare a boiler for an insurance inspection?

A. The fire should be all cleaned out the evening before the inspection is to take place; the damper should be opened to better facilitate the cooling down of the boiler. If the boiler is set in brick-work the water should not be blown out for at least two hours from the time the fire is drawn. If the inspection is to take place in the afternoon, it will be time enough to blow the water off in the morning; but if in the forenoon, the water will have to be let out some time during the night. Sweep out the tubes and connections to chimney; also sweep off all portions of the boiler shell that can be got at, so that the iron can be readily examined. If the boiler is of a return tubular or flue style, be sure and clean out the dust from combustion chamber. Now remove the man-hole and hand-hole plates, and wash or scrape out all of the sediment or deposit that may be found within the boiler. This will enable the inspector to get through in the shortest possible time.

Q. What is the difference in running a boiler that is insured or one that is not?

A. When we run an insured boiler, the inspectors are the ones to be obeyed in all things in connection with the boilers; but if the boilers are not insured, then the engineer should be the boss.

Q. If the water was discovered to be out of sight in the boiler, or if our water supply through any mishap were to be cast off from the boiler, what would be the best thing to do?

A. In case we wanted to suddenly do away with the heat under a boiler, shut the dampers and cover the fire with ashes, or fine coal, or anything of the kind that can be got at. This will cool the furnace down quicker than the old way of drawing the fire. After the fire is covered in this way, if it is found necessary it can be drawn.

Q. Should the engine be shut down in the above case?

A. If the engine is running, do not stop it; or if it is standing at the time, do not start it. This same rule refers to all steam and water valves in connection with the boiler.

Q. Is the loss very much from radiation of the exposed parts of boilers and pipes?

A. Yes; the loss is a great deal more than most engineers seem to be aware of.

Q. What is one of the best coverings for pipes and boilers to prevent loss by radiation?

A. Mineral wool is one of the best as well as the neatest covering made, though fossil meal and asbestos are often used as coverings to prevent loss by radiation.

Q. What are the best suggestions that can be given a new beginner about firing?

A. Fire evenly and regularly, a little at a time. Moderately thick fires are the most economical, but thin firing must be the order when the draught is poor. Take care to keep the grate evenly covered, and allow no air-holes in the fire. Do not "clean" fires oftener than is necessary.

Q. Why should the heating surface of a boiler be kept as clean as possible?

A. All heating surfaces of a boiler must be kept as free from soot and dust as possible, for they are such good non-conductors of heat that they hinder the water in a great measure from heating as soon as it otherwise would.

Q. How are boiler tubes usually made?

A. Boiler tubes are always lap-welded, and annealed at both ends; and when we speak of a tube of a certain size, we mean the external diameter.

Q. How are boilers tested?

A. For testing new boilers we usually give a cold water pressure of double the pressure of which we intend to run of steam; but after a boiler has been used we do not usually give a cold water test of more than one-half more than the steam pressure we wish to run.

Q. Of what does the hammer test consist?

A. The hammer test consists of a person going all over the surface of the boiler with a light hammer, striking light, even blows, so that if there is a thin place or a blister the operator will be able to detect it at once.

Q. What is a blister on a boiler?

A. A blister is a puffing out from the body of a boiler sheet where the layers of iron are not thoroughly welded in the process of manufacture.

Q. What is a bagging or a pocket on a boiler?

A. Bagging, bulging, or a pocket. are the various names given to a swelling on the fire-sheets of a boiler, generally found near the bridge wall on return tubular or flue boilers. It is different from the blister, as it includes the whole thickness of a sheet.

Q. How is the quickest way to remove vapor from a boiler after it has been blown off?

A. The quickest way to remove it is to open man-hole on top of boiler, then take out hand-hole plate in the front connection, shut furnace and ash-pit doors and open the damper; the draught of the chimney will draw it all out of the boiler. This rule, of course, refers to return tubular or flue boilers.

Q. Should the water be carried at a uniform height in a boiler?

A. Yes; the water should be carried at a uniform height in a boiler to avoid fluctuations in temperature and variations both in generation and pressure of steam.

Q. Is there any particular height that water should be carried in a boiler?

_A._ In an average-sized boiler of the return tubular or flue pattern, there should never be less than three inches of water above the top of the upper row of tubes.

_Q._ Suppose the engine stops suddenly from breakage, what must we do with the boiler?

_A._ Shut the throttle-valve, open the furnace doors, and close the dampers; bank the fires: and if this is not sufficient, pump up and blow off till the boiler is cool enough.

_Q._ What about cold and hot feed water for boilers?

_A._ Great advantage is gained by heating the feed water before it enters the boiler, as the chilling of the boiler plates is much reduced; and, further, the waste of heat, which means coal, is much reduced. Cold feed water, on entering a boiler, settles directly to the bottom, contracting the plates, causing deterioration of the metal, and often producing rupture of a serious nature. It is found that a continuous feed adds to the longevity of a boiler and facilitates the maintenance of steam at an even pressure.

_Q._ How can we tell whether water has been suffered to get too low in a boiler?

_A._ On examination of the inside of the boiler shell, a red coloration will show to about the point the water has fallen; and if there is any scale on the interior of the plates, it will be cracked off, so as to leave the iron perfectly clean down to the point to which the water has fallen.

_Q._ What effect does low water have on the tubes?

_A._ If the water has fallen low enough to bare the tubes, they will leak at the ends, and if there is any scale on them, it will be cracked off down to the point the water has reached.

_Q._ How do we proceed if a tube splits?

_A._ If dry pine plugs are driven into each end of the tube, it can be run till a new tube can be put in.

Q. Suppose a boiler plate cracks from one of the rivet-holes to the edge of the plate, how do you repair it?

A. A patch will have to be put on.

Q. Which is the most durable, large or small mud drums?

A. Small mud drums are the most durable.

Q. If you find cracks in the masonry of the boiler setting, what do you do, and why?

A. Point them up at once with cement, mortar or clay, to avoid chilling the boiler by the entering air-jets and loss of fuel by cooling the gasses.

Q. What is the proper thickness of fires with soft and hard coal?

A. With hard coal, from 4 to 7 inches; with soft coal, from 5 to 8 inches, depending on the coarseness of the fuel and the draught.

Q. How do you ascertain whether the boiler braces are all sound and tight?

A. By entering the boiler and inspecting them carefully.

Q. What is the effect of letting ashes accumulate in ash-pit?

A. Insufficient air supply for the fuel on grate and liability of burning out the grate bars.

Q. What is the effect of wetting down ashes and clinkers close to the boiler front?

A. Corrosion of the boiler metal, especially if it is an internally fired boiler.

Q. Is a " shaking grate " preferable to an ordinary grate bar?

A. Yes, by all means, in cases where it can be used.

Q. Give reasons why.

A. Because it obviates the necessity of opening the furnace doors so frequently, thus keeping the temperature of the furnace more uniform, and there is less loss of the fine coal dropping through the grate; it will save something in fuel; it will add to the power of a boiler, because the fire can be kept cleaner than with a stationary grate bar; and, last but not least, it saves a great deal in labor.

Q. Which is best, a spring loaded safety-valve or one with weight and lever?

A. A spring valve is the best, as well as the costliest; though a weight-and-lever is just as safe in the hands of a good engineer.

Q. How would we test the correctness of a safety-valve?

A. If a weight and lever, by figuring it out as per safety-valve rule; but if a spring valve, by having the steam gauge tested and running the pressure up to the blowing point on gauge, and seeing if the valve opens. The test gauge will prove the weight-and-lever valve just as well as the spring valve.

Q. Why is it advisable to scrape as well as to brush or blow out the tubes in a boiler?

A. A thin, carbonaceous scale is apt to form on the inside of the flues or tubes, which, although perhaps not thicker than a sheet of paper, requires the expenditure of considerable more fuel than if the surface was perfectly clean.

Q. What will cause external corrosion on a boiler?

A. Any dampness caused by the boiler, or pipes leaking, or the roof leaking upon the boiler, will cause corrosion.

Q. What action does the sulphur in coal have upon a boiler?

A. It forms an acid and corrodes the plates.

Q. What effect does water impregnated with sulphur have upon mud-drums when the feed-water is supplied through them?

A. It corrodes them badly.

Q. What are the principal impurities found in fresh water?

A. Carbonate of lime, magnesia, sulphate of lime, and salts of iron. Sometimes nitrates and chlorides are found.

Q. What are the mineral ingredients of sea-water?

A. Chloride of sodium, potassium and magnesium, bromide of magnesium, sulphates of lime and magnesia, and carbonate of lime.

Q. What effect upon the boiling point is caused when the water holds foreign substances in solution?

A. The temperature of the boiling point is raised; as, for instance, fresh water at the sea level boils at a temperature of 212° Fahr., while sea-water would boil at 213.3° Fahr.

Q. Suppose the pressure of steam in a boiler was suddenly lowered, what would be the result?

A. Steam would be rapidly disengaged from the water until the normal relations between the pressure and temperature were regained, both pressure and temperature being less than before the lowering of the pressure took place.

Q. What is " foaming," and what causes it in a boiler?

A. Foaming is violent ebullition in a boiler whereby water in small particles is thrown among the steam particles, mixing with them. It is caused by too little steam room; not sufficient surface at the water-level to allow the steam to disengage itself from the water quietly; crowding the tubes too closely in the boiler; from want of good circulation; from urging a boiler beyond its capacity; from oil, grease, or soda in the water; and other causes, such as changing from salt water to fresh, or the reverse.

Q. How can foaming be prevented or checked?

A. By feeding strongly and blowing out, by throttling

the engine a little, or by putting on a heavy fire to deaden down the heat. But in any case the boiler should be emptied and cleaned as soon as possible.

Q. What danger arises from it?

A. If the steam carries much water over the steam-pipe into the engine, there is danger of knocking out a cylinder head; or where steam is used to any extent for heating purposes, the water may be carried to such an extent out through the steam-pipes as to bring the water dangerously low in the boiler. And even a small amount of water carried over into the cylinder of the engine does a good deal of harm, for there will be small particles of grit in the water that will be carried over, and of course they will scratch the inside of the engine more or less.

Q. How are the joints of hand-hole and man-hole plates made?

A. With gaskets, usually of rubber, but sometimes of hemp.

Q. How do you prevent sticking of the gaskets?

A. By coating the gaskets on both sides with black lead (plumbago) and oil, or gum.

Q. What is the usual life of a land boiler?

A. From eighteen to twenty years.

Q. Of what is putty for joints made?

A. Generally of white lead, ground in oil, mixed with red lead to make the mass stiff enough to handle easily.

Q. Why do we cover the top of a boiler, pipes, etc.?

A. To prevent cooling of the metal, which would cause condensation of the steam and a consequent waste of fuel.

Q. Is a forced draught preferable to a natural draught, or one caused by the chimney alone?

A. Generally we consider a forced draught the best, because it can be kept the same all the time; whereas, if we depend on a natural draught, it will vary with the weather.

Q. What is a "reinforce" patch on a boiler?

A. It is a hard patch, circular in form, put on the outside of a boiler with counter-sunk rivets to give sufficient thickness to the boiler to allow a pipe to be screwed in. This is the way the lower blow-off pipe is always attached to the lower part of a boiler.

Q. What effect does large area of surface at the water line of a boiler have on the steam?

A. It gives dry steam and tends to prevent foaming.

Q. What effect does contracted steam room have on steam in a boiler?

A. It gives damp steam and tends to induce foaming.

Q. If a boiler is given to foaming, how can you prevent water from being carried over into the cylinder of the engine?

A. By carrying the water at a lower level, if possible.

Q. Should the safety-valve be fitted to the same outlet as the steam pipe?

A. No; it should be fitted with an independent connection.

Q. What will be the area of the chimney for a boiler?

A. From $\frac{1}{7}$ to $\frac{1}{16}$ of the area of the grate surface.

Q. How thick should the iron be in a flue of from 6 to 7 inches diameter?

A. Not less than .18 ($\frac{18}{100}$) of an inch for high pressure.

Q. How thick should the iron be in a flue from 8 to 9 inches diameter?

A. Not less than .2 ($\frac{2}{10}$) of an inch for high pressure or marine boilers.

Q. How thick should the iron be in a flue from 10 to 11 inches diameter?

A. Not less than .22 ($\frac{22}{100}$) of an inch for high pressure or marine boilers.

4

Q. How thick should the iron be in a flue from 12 to 13 inches diameter?

A. Not less than .23 ($\frac{23}{100}$) of an inch for high pressure.

Q. How thick should the iron be in a flue from 14 to 15 inches diameter?

A. Not less than .25 ($\frac{25}{100}$) of an inch for high pressure.

Q. How thick should the iron be in a flue from 15 to 16 inches diameter?

A. Not less than .27 ($\frac{27}{100}$) of an inch for high pressure.

Q. How much pressure would these flues stand with safety?

A. If in good condition, from 140 lbs. to 190 lbs. per square inch; the smaller one will stand the most.

Q. How do we determine the pressure per square inch allowable on lap-welded flues of not over 18 feet in length and from 7 to 16 inches in diameter?

A. Multiply the thickness of material in hundredths of an inch by the constant whole number 44, and divide the product by the radius of the diameter of the flue in inches; the quotient will be the pressure allowed.

Q. If we have a lap-welded flue of 14 inches diameter and .25 ($\frac{25}{100}$) of an inch thick, what pressure would be allowed on it?

A. Proceed thus :

.25 of an inch thick.
44 constant whole number.

100
100

Radius = 7)1100

157 lbs. pressure to the sq. in. allowed.

Always figure this way : consider the .25 just as if it was a whole number.

Q. How high does the Government inspector run the cold water pressure above the steam pressure that is intended to be run on a boiler?

A. The hydrostatic pressure is run up to one and one-half times the steam pressure to be run. Thus, if 100 lbs. of steam is to be run, the cold water pressure will be run up to 150 lbs.

Q. What is the greatest strain the Government allows to be put on braces, or stays, in her boilers?

A. Not more than six thousand (6,000) lbs. per square inch of cross sectional area; and they shall not be placed at a greater distance than 8.5 ($8\frac{1}{2}$) inches from center to center.

Q. How do we figure the bursting strength of boilers?

A. For a cylindrical boiler, made of either iron or steel plates, we multiply the tensile strength of the material in pounds by twice the thickness of the iron or steel in inches or parts of an inch, and divide the product by the diameter of the boiler in inches.

Q. Should all boilers be fitted with surface blow-offs?

A. Yes; and they are to be used when water is under violent ebullition.

Q. How do we know if a boiler is worked up to its full capacity?

A. If the boiler is running under a natural draught it is capable of consuming 12 lbs. of coal per each square foot of grate surface per hour.

SECTION VI.

ENGINES.

Q. What is the steam engine?

A. It is the motor through which steam transmits its power to machinery.

Q. How many classes of engines are there?

A. Two : throttling, or slide valve, and automatic cutoff engines.

Q. What is the principal difference in the two kinds?

A. The slide valve costs less. A man of less experience can run one, and the repairs, cost of packing, etc., will be less than the automatic. The automatic engine will pay for itself in from two to four years in the fuel it will save over a slide valve engine performing the same work.

Q. What is the rule for getting the horse-power (H. P.) of an engine?

A. Take the area of the piston in square inches and multiply it by the mean pressure per square inch piston, then multiply this product by the speed of the piston in feet per minute; now divide this last product by 33,000 and you will have the horse-power of your engine.

Q. How do we get the area of a piston in square inches?

A. Multiply the diameter of the piston in inches by itself, then multiply the product by .7854 ($\frac{7854}{10000}$); this will give the area in square inches.

Q. How do we get the mean pressure per square inch on the piston?

A. With an instrument called the "indicator;" but in a slide-valve engine running with a governor, the mean pressure will be about one-half of the lowest boiler pressure that will give the engine full speed.

Q. How do we get the speed of the piston in feet per minute?

A. Multiply the length of the stroke by two (because there are two strokes to each revolution), then multiply the product by the number of revolutions per minute the engine is making : this will be the speed of engine in feet per minute if the stroke was taken in feet; but if taken in inches the last product will have to be divided by 12 (because there are 12 inches per foot).

Q. What difference will there be in getting the H. P. of a condensing engine from a common high-pressure engine?

A. In the low-pressure or condensing engine there is a gauge to register the amount of vacuum. Take one-half of what this gauge registers and add it to the mean pressure if figuring by the slide-valve rule given above. If figuring with the indicator, the area of the card is figured the same as a high-pressure card.

Q. Get the H. P. of the following engine (slide-valve) : Diameter of piston, 12 inches ; length of stroke, 16 inches ; revolutions per minute, 140. The lowest boiler pressure that will give the engine full speed is 60 lbs.

A. Proceed thus :

```
 16  inches stroke.
  2  strokes per rev.
 ───
 32  inches per rev.
140  revs. per min.
 ───
1280
 32
 ───
12)4480   speed of piston in inches per min.
 ───
  373.3   "        "      feet   "    "
```

12 in., diam. of the piston.
12 "   "   "   "

—

24
12

—

144 circular inches, area of piston.
.7854 times circular inches = sq. in.

—

576
720
1152
1008

—

113.0976 sq. inches, area of the piston.
     30 lbs., mean pressure on piston.

—

3392.928 lbs., total pressure on piston.
    373.3 speed of piston in feet per min.

—

10178784
10178784
23750496
10178784

—

Ft. lbs. }
per H.P. } 33000)1266580.0224(38.3 + H.P. is what this engine is capa-
     99000   ble of generating.

—

276580
264000

—

125800
99000

Q. Give good proportions of an 80-H. P. engine of the slide-valve pattern.

A. Diameter of the piston, 15 inches.

Length of the stroke, 20 inches.

Sq. in. area of the piston, 177+.

Revolutions per minute, 172.

Speed of the piston, 573+ feet per minute.

Mean pressure on piston, 26 lbs.

The cut-off of steam takes place at about ¾ that of the stroke.

Length of the crank, 10 inches.

Length of the connecting rod, 50 inches.

Diameter of the piston rod, 2 inches.

Diameter of the valve rod, 1¼ inches.

Diameter of the steam pipe, 3¾ inches.

Diameter of the exhaust pipe, 4½ inches.

Area of steam ports, each, 10¼ square inches.
Area of exhaust port, 24½ square inches.
Area of steam pipe, 11½ square inches.
Area of exhaust pipe, 14¼ square inches.

Q. How is the speed of an engine governed?

A. The speed of an engine is regulated by the governor. In the throttling engine it throttles the steam in the supply pipe, but in the automatic engine it regulates the admission valves in the quantity of steam they admit to the cylinder.

Q. What is meant by terminal pressure?

A. By terminal pressure we mean the pressure of steam in the cylinder just as it is released.

Q. What is meant by mean pressure?

A. The average pressure per square inch throughout the stroke.

Q. What is meant by initial pressure?

A. It is the pressure of steam as it enters the cylinder.

Q. What is meant by gauge pressure?

A. It is pressure per square inch above the atmospheric pressure.

Q. How could the terminal pressure of steam be found?

A. Multiply the initial pressure by the number of inches the piston has travelled when the cut-off of steam takes place, and divide the product by that portion of the stroke in inches at which the release or exhaust valve opens. The quotient will be the terminal pressure.

Q. What three things give us power in an engine?

A. Size, pressure and speed.

Q. Upon what does the steam act to give us power in the steam engine?

A. The steam is admitted to the cylinder between the piston and cylinder head, the head being bolted on to the cylinder firmly, of course. The piston is the part that is driven with the force of the steam to the opposite end of the cylinder. There the operation is repeated, and the piston is driven back to its former place.

Q. What is the length of the crank on an engine?

A. The length of the crank is just one-half of the length of the stroke of the engine.

Q. How many strokes does an engine or pump make for each revolution?

A. Two strokes per revolution.

Q. What should be the diameter of the steam supply pipe for the engine?

A. About one-fourth the diameter of the piston.

Q. How much larger should the exhaust pipe be than the steam supply pipe for the engine?

A. The exhaust pipe for letting the steam out of the engine after it has done its work should be one-fifth larger than the steam pipe.

Q. In a slide-valve engine what will be the diameter of the valve rod?

A. About one-twelfth of the diameter of the piston.

Q. What will be the diameter of the piston rod?

A. In a slide-valve engine, usually about one-eighth of the piston diameter; but for high pressure, such as we usually get in automatic engines, we have a special rule which will appear later.

Q. What are steam ports in a steam engine?

A. Steam ports are the passages through which the steam is admitted from the steam chest to the cylinder.

Q. How do you find the area or size that these steam ports should be to admit steam enough to the cylinder?

A. Take the area of the piston in square inches and divide it by 19; this will give you the area of each steam port in square inches if the speed of the piston is not to exceed 500 feet per minute. If the speed is greater than

500 feet per minute, then we would take the area of the piston in square inches and multiply it by the speed of the piston in feet per minute, and divide the product by 10,000; this will give the area of each steam port in square inches.

Q. What is meant by the exhaust port?

A. It is the passage through which the steam comes back from the cylinder after doing its work there.

Q. How do we get the area of the exhaust port?

A. For a slide-valve engine with one central exhaust port they usually have the area of it about one-eighth of the area of the piston, or one-fifth larger than the area of both steam ports combined.

Q. What are the five principal places to oil on any engine?

A. The inside of the steam chest and the cylinder, the two main bearings on the crank shaft, the crank pin, and the eccentric.

Q. What four things are most apt to make our engine heat up her bearings?

A. Our main belt being very taut, keying up or tightening up too much, dust or dirt of any kind getting into the bearings, or if our engine is in a very hot room.

Q. Which is the cheaper, a leather or a rubber belt for a main belt from our engine?

A. In most cases a rubber belt will be the cheaper.

Q. How many kinds of oil do we use on an engine?

A. Two: cylinder oil, which will stand from 400° to 600° of heat, for the inside of steam chest and cylinder, and common machinery oil for all external parts.

Q. What is a good length for a connecting rod (the rod which connects the cross-head to the crank) for an engine?

A. About five times the length of the crank, center to center.

Q. When is the best time to key up or tighten up the bearings on an engine?

A. When the engine is warm, as it will be after running a few hours, for then the pins, journals, etc., have had sufficient chance to expand all that they will with the maximum running temperature.

Q. How is the best way to oil an engine?

A. Put on a little at a time, and often.

Q. What are some of the principal dangers an engine is subjected to?

A. Water getting too high in the boilers and coming over into the steam pipe, and from there into the cylinder, is liable to knock out a cylinder head; bearings getting heated; nuts or keys working loose.

Q. What are some of the most important things for an engineer to look after?

A. To see that the steam does not get too high; to see that the water does not get too low; to see that the water does not get too high; to see that the bearings do not get hot, nor anything work loose about the engine; to see that the steam does not get too low; and, last of all, keep everything clean in and about the engine and boiler room. These are the engineer's duties in their order of importance.

Q. What is another difference between the automatic and the slide-valve engines?

A. In the slide-valve engine one valve performs the four separate operations of letting steam alternately into both ends of the cylinder and out of it, while in the automatic engine there are four separate valves to perform the same work.

Q. Do we know where the steam is cut off in an automatic engine?

A. We do not as a general thing; for if the work varies, the cut-off will vary.

Q. What is the difference in the amount of water consumed with slide-valve and automatic engines?

A. Automatic engines use from 20 to 30 lbs. of water per H. P. per hour, while slide-valve engines use from 40 to 60 lbs. of water per H. P. per hour.

Q. What is the difference in the consumption of coal in the two engines?

A. The Corliss valve gear was the first automatic engine made, and is considered about the average today; it will consume about $2\frac{1}{2}$ lbs. of coal per H. P. per hour. The average slide-valve engine will consume from $3\frac{1}{2}$ to 4 lbs. of coal per H. P. per hour.

Q. What is the extreme variation of coal consumption of the different engines used today?

A. From 2 to 5 lbs. of coal per H. P. per hour.

Q. How do we find the proper length of steam ports on the cylinder side of a slide-valve engine?

A. Divide the diameter of the cylinder by 1.2 ($1\frac{2}{10}$); the quotient will be the length required.

Q. If we have the area of a piston given, which is the quickest way to find the area of the steam supply and the exhaust pipes?

A. Divide the area of the piston by 16; the quotient will be the area of the steam supply pipe. And if we divide the area of the piston by 13, the quotient will be the area the exhaust pipe ought to be.

Q. How do we find the area of a piston rod where high-pressure steam is to be run, as in the automatic engine?

A. The proper way to find the area of a piston rod in an automatic engine is to multiply the area of the piston in square inches by the highest boiler pressure to be run, and divide the product by 4,480; nine-tenths of the quotient will be the area in square inches that the piston rod should be if made of steel.

Q. What is a good proportion between diameter of cylinder and the length of stroke of an engine?

A. For most work the proportion of the diameter of the cylinder to the length of stroke is as 3 to 4. The well-known Corliss has a proportion of 1 to 2; but some builders (as the Armington & Sims Mfrs.) construct their engines " square, " that is, the diameter of cylinder is the same as the length of stroke.

Q. How do we set a slide valve?

A. To set a slide valve is simply to lengthen or shorten the distance between the eccentric and valve, or, as some authorities say, to equalize the vibrations of the valve. It does not make any difference whether the eccentric is set or not. In setting a valve, all we have to do is to see that it is made fast to the shaft and that all the connections are tight enough for running order.

Q. How do we set an eccentric?

A. Place the engine on one of the dead centers. Now turn the eccentric in the direction we wish the engine to run until the part corresponding to the end the piston is at commences to open; make the eccentric fast to the shaft; then turn the engine round to the other dead center and see if the valve has the same amount of lead or opening there. For an engine of ordinary speed (from 300 to 500 feet per minute), give the valve about $\frac{1}{32}$ of an inch lead. If the engine is run faster, give more lead.

Q. In a new engine, which has to be set first, the valve or eccentric?

A. The valve must always be set first in the slide-valve engine, for the eccentric could not be set until the valve had been.

Q. What is meant by lead and lap of a valve?

A. The distance that the steam ports are open when the engine is on the dead center is called the lead, and the distance the valve reaches beyond the steam ports at either end, when the eccentric is at one-half stroke, is called the lap of the valve.

Q. About how many living horses is a steam H. P. equal to?

A. A steam H. P. is equal to about three average horses' power. An average horse is equal to about seven average men in power.

Q. Is there any other standard for measuring the power of engines?

A. There are only two systems of measuring the power of engines used in Europe or America, viz. : the English standard, fixed by James Watt, which is 33,000 lbs. raised one foot high in one minute's time, equals one H. P. ; and the French, which is called the *force du cheval ;* this is based on the metric system, and is 4,500 kilograms one meter high in one minute's time. This will be about $\frac{1}{73}$ less than our standard.

SECTION VII.

BELTS, SHAFTING, SPEED.

Q. How does the power of belts increase?

A. The power that a belt can transmit varies directly as its width and speed (if all other conditions remain the same) with a limit of 5,000 to 6,000 feet per minute; that is, this has been found to be about the greatest speed advisable to run belts.

Q. What other conditions are there to be considered?

A. There are two other things to be considered: first, the thickness, and next, the tautness of the belts.

Q. What is it gives us power in belts?

A. Friction is what enables us to transmit power with belts.

Q. What three things give us the friction in belts?

A. Surface the belt covers, speed that it travels and the pressure that is brought to bear upon it.

Q. How fast will a one-inch single leather, or a three-ply rubber, belt have to travel to be able to transmit one H. P.?

A. A one-inch single leather or a three-ply rubber (which are of about the same thickness and weight) would have to travel 800 feet per minute to transmit one H. P., or 1,600 feet per minute to transmit two H. P., and 2,400 feet per minute to transmit three H. P., and so on; this is running them as taut as they should be run.

Q. How fast will a two-inch belt have to run to be able to transmit one H. P.?

A. A two-inch single leather or a three-ply rubber belt will have to run 400 feet per minute to transmit one H. P.; 800 feet per minute to transmit two H. P.; 1,200 feet per minute to transmit three H. P., and so on.

Q. How many different thicknesses of belts are there?

A. Leather belts come in two thicknesses, single and double; rubber belts come in four thicknesses, three-ply, four-ply, five-ply and six-ply.

Q. How does the strength of shafts increase?

A. The strength of shafts, for either a binding or a twisting strain, varies as their speed and the cube of their diameters; for a two-inch shaft will do eight times as much work as a one-inch shaft running the same speed.

Q. How fast will a one-inch shaft have to run to transmit one H. P.?

A. A one-inch shaft running 100 revolutions per minute will transmit one H. P.; 200 revolutions per minute, two H. P.; 300 revolutions per minute, three H. P., and so on.

Q. How much strain would a one-inch shaft twelve feet long stand at the end of a crank one foot long?

A. It would safely stand 50 pounds.

Q. How fast will a two-inch shaft have to run to transmit one H. P.?

A. A two-inch shaft running 12½ revolutions per minute will transmit one H. P.; 25 revolutions per minute, two H. P.; 50 revolutions per minute, four H. P., and so on.

Q. How many H. P. will a three-inch shaft transmit, running 100 revolutions per minute?

A. Twenty-seven H. P., and if it run 200 revolutions per minute it would transmit 54 H. P., and so on.

Q. How do we find the maximum H. P. of a shaft within good working limits?

A. Multiply the cube of the diameter of the shaft in inches by the speed in turns per minute, and divide by 205 if a cast-iron shaft, or by 110 if wrought iron, or by 82 if a steel shaft. The quotient is the H. P. of shaft.

Q. How do we find the diameter of a shaft capable of transmitting a given H. P. within good working limits?

A. Multiply the H. P. by 205 if a cast-iron shaft, or by 110 if wrought iron, or by 82 if a steel shaft, and divide by the speed in turns per minute. The cube root of the quotient is the diameter in inches.

Q. How do we find the speed required to run a shaft, for transmitting a given H. P. within good working limits?

A. Multiply the H. P. by 205 if a cast-iron shaft, or by 110 if a wrought iron shaft, or by 82 if a steel shaft, and divide the product by the cube of the diameter of the shaft in inches. The quotient is the speed in turns per minute.

Q. How do we find the size of a driving pulley to give another shaft a given speed, where speed of driving and driven shafts are known as well as the size of driven pulley?

A. Multiply the diameter of the driven by the number of revolutions per minute it is running, and divide the product by the revolutions of the driving shaft. The quotient will be the diameter of the driving pulley.

Q. How do we find the size of a driven pulley?

A. Multiply the diameter of the driving pulley by its number of revolutions per minute, and divide the product by the number of revolutions per minute the driven shaft is to run. The quotient will be the diameter of the driven pulley.

Q. How do we find the speed of a driven shaft, where speed of driving shaft is known, as well as size of both driving and driven pulley?

A. Multiply the diameter of the driving pulley by the number of revolutions it runs per minute, and divide the product by the diameter of the driven pulley. The quotient will be the number of revolutions the driven shaft makes per minute.

Q. How do we find the relative amount of centrifugal force of different pulleys?

A. If there be two wheels of the same weight, and making the same number of revolutions per minute, but the diameter of one be double that of the other, the larger will have double the amount of centrifugal force; or if the velocity of a wheel be doubled it will have four times the amount of force.

Q. How many different rules have we for finding the area of circles?

A. Five, as follows: Multiply the circumference by $\frac{1}{4}$ of the diameter; multiply the square of diameter by .7854 ($\frac{7854}{10000}$); multiply the circumference by .07958 ($\frac{7958}{100000}$); multiply $\frac{1}{2}$ of the circumference by $\frac{1}{2}$ of the diameter; and multiply square of the radius ($\frac{1}{2}$ diam.) by 3.1416 ($3\frac{1416}{10000}$).

Q. What do we know about circles?

A. Diameter of a circle multiplied by 3.1416, the product equals the circumference; circumference divided by 3.1416, the quotient equals the diameter; diameter multiplied by .8862 ($\frac{8862}{10000}$), the product equals side of a square of equal area; and a side of a square divided by .8862, the quotient equals diameter of a circle of equal area.

Q. What do we know about a cubic foot?

A. A cubic foot equals 1,728 cubic inches; 2,200.15 cylindrical inches; 3,300.23 spherical inches; 6,600.45 conical inches; 62.32 lbs. of fresh water at a mean temperature; 7.48 United States standard gallons; 452 lbs. cast iron; 485 lbs. wrought iron, and 489 lbs. of steel.

Q. How many different thicknesses of steam pipe are there?

A. Three: the common, which is the lightest; the extra, and the double-extra, which is the thickest; external diameters are all same size.

Q. How are the internal diameters?

A. The size of steam pipe goes by the internal diameter, as for instance, a one-inch extra steam pipe is just

5

one inch inside diameter; the one-inch common steam pipe is a little more than an inch inside diameter, and one-inch double-extra is not quite one inch inside diameter.

Q. How long does the steam pipe come?

A. The regular length is 16 feet.

Q. What is the smallest and largest wrought iron steam pipe made?

A. The smallest steam pipe made is $\frac{1}{8}$ of an inch and the largest is 24 inches diameter, but it has been proposed to make it of even 42 inches diameter.

Q. How many different ways have we of welding pipe together?

A. Two, the " lap weld" and the " but weld." Boiler tubes are always "lap welded," and unlike the steam pipes they are rated by their external diameter.

Q. What is a good receipt for a solder?

A. Coarse plumbers' solder contains lead, two parts; tin, one part; common solder contains equal parts of the two metals; fine solder is composed of two parts of tin and one part of lead.

Q. What is a good thing for cleaning brass-work on engines and boilers?

A. There is nothing better than oxalic acid and salt water, in the proportion of $\frac{1}{2}$ an ounce of acid to a pint of salt water.

Q. What of the common metals expands the most?

A. Zinc expands the most, copper the next and brass, of course, being composed principally of copper, expands nearly as much as the copper.

Q. How is a good way of measuring belting in the roll?

A. Take the sum of the inside and outside diameters in inches and multiply it by the number of turns made by the belt (in the roll), then multiply the product by .1309 ($\frac{1309}{10000}$); this will give the length of the belt in feet.

PART II.

SECTION I.

COMBUSTION OF COAL

CHEMICALLY CONSIDERED.

In any attempt to ascertain how to develop the largest percentage of real paying duty from a given amount of fuel — the most heat and power — our first inquiry should be as to the nature of the fuel and the elements of combustion that enter into it — considered from a chemical standpoint.

From scientific analyses by Professor Liebeg and others, it is shown that in the various kinds of *soft* or bituminous coal there is about 80 per cent of *carbon*, 5 per cent of *hydrogen*, 10 per cent of azote and *oxygen* and 5 per cent of *ash;* these proportions varying somewhat in the different kinds. The principal constituents of all coal, however, are *carbon* and *hydrogen*, which are united and solid in its natural state. In all bituminous coal *hydrogen* is the main element from which *gas* is evolved, and by the combustion of which *flame* is produced.

Their main constituents, *carbon* and *hydrogen*, united in the solid coal are essentially different in character and in their modes of entering into combustion, and to the ignorance or neglect of this primary distinction, much of the waste and uncertainty attending the use of coal on a large scale is due. The *theory* of combustion is

well understood by scientific men, yet *practically* the art
of burning coal economically remains at a low ebb, and
the science of converting the natural elements of coal
into heat and power is but little understood. While it is
well known that the constituents of coal — carbon and
hydrogen — require certain quantities of atmospheric
air to effect their combustion, yet *practically* the means
necessary to ascertain *what* quantity is supplied is often
neglected and the matter is generally treated as though
the *right* proportion was unimportant. While theoretically the relative constituents of which atmospheric air
is composed are generally well known, yet practically the
nature of these constituents or their effects in combustion is often totally ignored. It is known scientifically
that the inflammable gases are combustible and convertible into heat, *only in proportion to the right mixture and
union effected* between them and the *oxygen of the air;*
yet in practice we are not apt to trouble ourselves as to
whether such mixture is effected or not. These and
similar illustrations indicate a lack of practical knowledge of how best to utilize the natural elements of coal
so as to produce heat and power economically.

The bituminous portion of coal is convertible to heat
in the *gaseous state alone*, while the carbonaceous portion on the contrary is combustible *only in the solid*
state, and *neither can be consumed while they remain
united;* hence to effect combustion their separation must
be effected and a *new union* formed with other elements,
viz., atmospheric air or oxygen. In combustion, there
must be a *combustible* and a *supporter* (oxygen) of combustion, which means *chemical union.*

Until all the *bituminous* constituents are evolved from
coal its solid or carbonaceous part remains black at a
comparatively low temperature, and inoperative as a
heating body, and must wait for the heat essential to its
combustion in its own peculiar way. And if the bitumi-

nous part be not consumed or utilized, it would be
better were it not in the coal, in which case such heat
would be saved and available for duty. To this fact
may be attributed the alleged greater heating properties
of anthracite coal or coke over bituminous coal.

Having noticed the leading properties of coal in its
natural state and the elementary divisions, bituminous
and carbonaceous, the next important consideration is
its union with the atmospheric air; and it will be found
that the *practical* economy in the use of coal is intimately
connected with the combustion of the gases evolved
from the coal, which cannot be effected without a *suitable
mixture* with the air, the principal supporter of combustion.

Fresh coal supplied to the glowing coals in the furnace
does not immediately or instantly increase the general
temperature, but becomes an absorbent of it and the
source of the volatization of the bituminous portion of
the coal (of the generation of gases); and volatization is
the most cooling process of nature. by reason of the
quantity of heat directly converted from the "sensible"
to the "latent" state.

On the application of heat to bituminous coal, the first
result is its absorption by the coal and the disengage-
ment or liberation of the gases, from which flame is
exclusively derived. The constituents of this gas are
hydrogen and *carbon;* and the union is called *carburetted
hydrogen* and *bi-carburetted hydrogen*, commonly called
olefiant gas. This *carburetted hydrogen* though not a
combustible — taken by itself — becomes a combustible
when *united with* oxygen (and for this reason oxygen is
called a "supporter"), *neither of which*, however, *taken
alone can* be consumed. Coal gas, whether generated in
a retort or furnace, is essentially the same. By itself it
is not inflammable, can neither produce flame nor permit

its continuance in other bodies. A lighted taper is instantly extinguished, if introduced into a tank of hydrogen gas. In short, for all practical purposes combustion is more a question of *air* than *gas*.

In the common gas-burner we have an illustration of our ability to control the gas for light and heat. In the furnace, however, we have no control over the gases, as to quantity, after throwing on the coal, but can exercise considerable control over the *air*, in all essentials to perfect combustion. This control of the air is what has brought the lamp to such perfection, and may be made equally available for the furnace. The difficulty in the way is, that it has to be controlled on a much larger scale, but as regards quantity and quality, the principle is the same. How, when and where this controlling influence over the admission of air is to be exercised to the best advantage in burning fuel in a furnace, are questions demanding the most careful attention of the practical engineer, and they must be decided on *strict chemical principles.**

The first essential in effecting the combustion of gas is to ascertain the quantity of *oxygen* with which it will chemically combine; next, the quantity of *air* required to supply the necessary quantity of oxygen. Now while this may be scientifically understood and correctly arrived at by an expert chemist in the laboratory, it,

---

* In ordinary language, a body is said to burn when its elements unite with the oxygen of the air and form new products. One of the bodies, as hydrogen, is termed the burning or combustible body, and the oxygen is said to be the supporter of combustion; but this language, although convenient for common use, is incorrect as a scientific expression, for oxygen may be burned in a vessel of hydrogen, as well as hydrogen in a vessel of oxygen; the one and the other being equally active in the process, and being related to each other in every way alike. — *Elements of Chemistry, by* ROBERT KANE, M.D.

however, is to be expected that, in the management of combustion in a furnace, the ordinary engineer can at best only approximately apply the exact laws of chemistry to the very imperfect conditions found at every furnace. It is important, however, that every engineer should, at least, understand theoretically the analysis of the elements he has to deal with in producing combustion, and the proportional part of each element entering into the same; therefore, a brief treatment of this subject from a chemical point of view is here presented, selected from the writings of the best authorities on this subject.

According to chemical analysis, an atom of *hydrogen is double* the bulk of an *atom of carbon vapor;* yet the latter is *six times the weight* of the former. Again, an atom of *hydrogen* is *double* the bulk of an atom of *oxygen;* yet the latter is *eight* times the weight of the former. So of the constituents of atmospheric air, nitrogen and oxygen. An atom of the former is double the bulk of an atom of the latter; yet in weight, it is as fourteen to eight. It is also ascertained that *oxygen* is but *one-fifth* of the bulk of *air*. Five volumes of the latter (air) will necessarily be required to produce *one* of the former (oxygen). And, as we want *two* volumes of oxygen for each volume of the coal gas, it follows that, to *obtain those two volumes*, we must provide *ten volumes of air*.

Thus far the subject has been treated mainly with reference to the supply of *air* required for the saturation and combination of the *gaseous* portion of coal. We will now consider a corresponding question with reference to the *carbonaceous* part of coal. which rests upon the grate-bars in a solid form, after the gaseous matter has been evolved therefrom.

It is stated by chemists, that *carbon* is susceptible of uniting with oxygen in three proportions, by which *three distinct bodies* are formed, possessing *distinct chemical properties*. Although this peculiarity of the *union* of *carbon* with oxygen is almost wholly neglected in practice, yet it is very essential in correctly estimating the *quantity of air* which should be admitted to the furnace.

These proportions, in which carbon unites with oxygen, form: 1. *Carbonic acid;* 2. *Carbonic oxide;* and 3. *Carbonous* or *oxalic acid.* With the first and second, we have to deal principally in the furnace. Carbonic *acid* is a compound of *one* atom of carbon with *two* atoms of oxygen; while carbonic *oxide* is composed of the same quantity of carbon with but *half* the above quantity of oxygen. Here we see that carbonic oxide, though containing but one-half the quantity of oxygen, is yet of the same volume as carbonic acid, which is of considerable importance on the question of *draught* and supply of *air.* Now the combustion of this *oxide* by its conversion into the *acid,* is as distinct an operation as the combustion of the carburetted hydrogen, or any other combustible; yet all this is almost entirely overlooked in practice at the furnace.

Outside of the laboratory, and in actual practice, but little is known as regards the *formation of this oxide.* The *direct effect* of the union of *carbon and oxygen* is the formation of *carbonic acid.* If, however, we abstract one of its portions of oxygen, the remaining proportions would then be those of carbonic oxide. It is equally clear that if we *add* a second portion of *carbon* to carbonic acid, we shall arrive at the same result, namely, have carbon and oxygen combined in equal proportions, as we have in carbonic oxide. By the addition of still another portion of carbon *two* volumes of carbonic *oxide* will be formed. Now, if these *two* volumes of carbonic *oxide* cannot find the *oxygen* required to complete their *saturating* equivalents, they pass away necessarily but *half con-*

sumed — a process which is constantly going on in all furnaces, where all the air has to pass through a body of incandescent carbonaceous matter.

This frequently leads to a common error in what is called the "combustion of smoke"; for if the carbonaceous constituent of coal, while yet at a high temperature, encounters carbonic *acid*, this latter, taking up an additional portion of carbon, is converted into carbonic *oxide* and again becomes a gaseous and invisible combustible. The most prevailing operation of the furnace, however, by which the largest quantity of carbon is lost in shape of carbonic *oxide*, is thus : The air, on entering from the ash-pit, gives out its oxygen to the glowing carbon on the bars, and generates much heat in the formation of carbonic acid. This *acid*, necessarily at a very high temperature, passing upwards through the body of incandescent solid matter, takes up an additional portion of carbon, and becomes carbonic *oxide*.*

Thus by the conversion of one volume of *acid* into two volumes of *oxide*, heat is actually absorbed, while the carbon taken up during such conversion is also lost, and we are liable to be deceived by imagining we have "burned the smoke."

The formation of this compound, carbonic oxide, being thus attended by circumstances of a curious and involved nature, is probably the cause of the prevailing ignorance of its properties ; for, while we find everywhere the term *carbonic acid*, as a product of combustion, we hear

---

* "Carbonic oxide may be obtained by transmitting carbonic acid over red-hot fragments of charcoal contained in an iron or porcelain tube. It is easily kindled; combines with half its volume of oxygen, forming carbonic acid, which retains the original volume of carbonic oxide. The combustion is often witnessed in a coke or charcoal fire. The carbonic acid produced in a lower part of the fire is converted into carbonic oxide as it passes up through the red hot embers." — *Graham's Elements of Chemistry.*

nothing of *carbonic oxide* — one of the most waste-induc-
ing compounds of the furnace, unless provided with its
equivalent volume of air, by which its combustion will
be affected.*

Another important peculiarity of this gas (carbonic
oxide) is that, by reason of its already possessing *one-
half* its equivalent of oxygen, it inflames at a lower
temperature than the ordinary coal-gas; the consequence
of which is, that the *latter*, on passing into the flues, is
often cooled down below the temperature of ignition;
while the *former* is sufficiently heated, even after having
reached the top of the chimney, and is there ignited on
meeting the air. This is the cause of the red flame often
seen at the top of chimneys, or the funnels of steam-
ships.

We may thus conclude for a certainty, that, if the
carbon, either of the gas or of the solid mass on the
bars, passes away in union with oxygen in any other
form or proportion than that of *carbonic acid*, a proportionate loss of heating effect is the result.

According to chemical analysis *ten* cubic feet of *air* is
required to supply *two* cubic feet of oxygen to effect the
combustion of *one* cubic foot of coal-gas; but if this
quantity of air does not contain this 20 per cent, it is

---

* "Among the stove-doctors of the present day, none are more
dangerous than those who, on the pretense of economy and con-
venience recommend to keep a large body of coke burning slowly,
with a slow circulation of air. An acquaintance with chemical
science would teach them, that, in the obscure combustion of coke
or charcoal, much carbonic oxide is generated, and much fuel consumed, with the production of little heat; and physical science
would teach them, that, when the chimney draught is languid,
the burned air is apt to regurgitate through every seam or crevice,
with the imminent risk of causing asphyxia, or death, to the in-
mates of apartments so preposterously heated." — *Dr. Ure's Paper
on Ventilating and Heating.* (Read before the Royal Society.)

clear we cannot obtain the requisite amount. Therefore, when we speak of mixing a given quantity of oxygen with a given quantity of coal-gas, it is because we know that the former is required to saturate the latter; so when we speak of mixing a given volume of atmospheric air with a given volume of coal-gas, we do so knowing that such precise quantity of *air* will provide the requisite quantity of oxygen. If, however, by any means, the air employed has either lost any portion of its oxygen, or is mixed with any other gas or matter, it no longer bears the character of pure atmospheric air, and cannot satisfy the condition required as to *quantity of oxygen* which is essential. The air in such cases may be said to be deteriorated or vitiated, and therefore the *quality* of the air employed is entitled to consideration.

Having considered the *constituents of coal*; the *necessary conditions* to perfect combustion; the nature of the *gases evolved* from coal; the proper *mixture* of the *gases with oxygen* — the promoter of combustion — and the quantity and quality of the air employed, let us now inquire how far the usual methods of constructing and managing furnaces satisfy those conditions.

First, it should be noticed, that the conditions required vary somewhat in the different kinds of fuel, especially in anthracite and bituminous coal; there being in the latter larger quantities of hydrogen gas to utilize, with which a separate and suitable quantity of air or oxygen must be mixed to get the best results in the way of heat and power.

The two distinct operations of supplying air to the *gas* generated in the upper part of the furnace, and to the *carbon* resting on the grate-bars, is not sufficiently considered in daily practice. By the usual method, the whole supply of air is compelled to pass through the ash-pit, and the solid carbon upon the bars; the

air which has already been employed in a separate
and destructive process is thus brought to the gases,
insufficient in quantity and quality; and it is expected
that the result will be satisfactory and combustion complete; then, when it is found that, instead of producing
carbonic acid and water, we have produced a large
volume of *smoke* or unconsumed combustible matter, we
set about inventing some process by which this "smoke"
is to be consumed.

There are a number of so-called " smoke-consumers."
It is, however, no part of the writer's purpose to show
how smoke can be burned, but to show the conditions
necessary to *burn coal without smoke*. On the contrary,
we contend that, when smoke is once produced in a
furnace, it is as impossible to burn it or to convert
it into heat as it is to convert the smoke from the flame
of a candle or lamp into light and heat. In short, without controlling the air, but poor results are obtained in
either case. For an illustration, take an Argand burner,
poorly or improperly adjusted, as regards the air, and
we get a smoky, murky light; properly adjusted, as regards the air, we get a white, clear light and heat. In
the latter case, would it be correct to say, the "lamp
burns its smoke," or the "lamp burns without smoke?"

Knowing the quantity and quality of air to be admitted
to the furnace, the next important consideration is the
discovery of the *best method of such a mixture* of the *air
and gases* required for complete combustion. There have
been many mechanical devices for supplying the gases
above the coal in the combustion-chamber with fresh air,
each possessing more or less merit, but in this direction
there is still room for great improvement; and it will be
found that any mechanical device, that does not carefully
take into account the principles of chemistry, will never
accomplish the desired result.

In common practice it is taken too much for granted, that, if air by any means be introduced to the furnace, it will, as a matter of course, mix with the gases liberated from the coal or other combustibles, whatever be the nature or state of such fuel. It is, however, the proper distribution of the air, and the bringing together bodies of gas and air in a state of preparation, so that the *requisite mixture of all the elements* will be incorporated and utilized, that effects perfect combustion.

It should be observed, that in effecting these mixtures of gases, they will properly combine and become united, if *sufficient time be allowed;* but in the furnace, as it is impossible to *force* the gas and air to mingle with sufficient rapidity under ordinary circumstances, our attention should be directed towards making such modifications of the furnace as will aid nature in those operations essential to combustion.

Prof. Daniels says, " There can be no doubt that the affinity of hydrogen for oxygen, under most circumstances, is stronger than that of carbon." He further says, " With regard to the different forms of hydrocarbon, it is well known, that the whole of carbon is never combined with oxygen in process of detonation or silent combustion, unless *a large excess of oxygen be present.*"

" For the complete combustion of olefiant gas, it is necessary to mix the gas with *five times* its volume of oxygen : *three only are consumed.* If less be used, part of the carbon *escapes combination,* and is deposited as a black powder. It is clear, therefore, that the whole of the hydrogen of any of these compounds of carbon may be combined with oxygen, while a part of their carbon may escape combustion, even when enough oxygen is present for its saturation."

" That which takes place when the mixture is designedly made in the most perfect manner must, undoubtedly, arise in the common processes of combustion, where the mixture is fortuitous and much less intimate. Any method of insuring the complete combustion of fuel, consisting partly of the volatile hydro-carbon, *must be founded upon the principle of producing an intimate mixture with them, of atmospheric air, in excess,* in that part of the furnace to which they naturally arise. In the common construction of furnaces, this is scarcely possible, as *the oxygen of the air, which passes through the fire-bars, is mostly expended upon the solid part of the ignited fuel with which it first comes in contact.*"

In view of the foregoing opinions from Prof. Daniels, it is clear, that some better device is needed at furnaces than we have yet had, by which air or oxygen may be taken into the combustion-chamber, as well as through the grate-bars or fire, especially for bituminous coal, if we would utilize the gaseous elements of the fuel; and any device that will effect the most intimate mixture of these gases with the oxygen of the air, will be most successful in securing *perfect combustion.*

We have many convincing illustrations of what nature requires, showing the importance of bringing air to the gases to effect perfect combustion, of which the common candle and Argand burner lamp are fair samples.

Mr. Brande observes, "In a common candle the tallow is drawn into the wick by capillary attraction, and there converted into vapor, which ascends in the form of a conical column and has a temperature sufficiently elevated to cause it to combine with oxygen of the surrounding atmosphere, with a temperature equivalent to a *white heat.* But this combustion is *superficial only,* the flame being a thin film of white-hot vapor, enclosing an interior portion *which cannot burn for want of oxygen.* It is in consequence of this structure of the flame that we so materially *increase its heat*, by propelling a current of air through it by means of the blow-pipe."

Dr. Reid says, "The flame of the candle is produced by the gas formed around the wick acting upon the oxygen of the air; *the flame is solely at the exterior* portion of the ascending gas. All *without* is merely heated air, or the products of combustion; all *within is unconsumed gas*, rising in its turn to affect (mingle with) the oxygen of the air."

Berthier, Vol. I., p. 177, observes, "The flame presents four distinct parts, namely: 1. The *base*, of a sombre blue; this is the gas that burns with difficulty, because it has not yet acquired a sufficiently high temperature. 2. An *interior dark cone;* this is combustible gas *highly heated, but which does not burn, because it is not mixed with air*. 3. The *brilliant conical envelope;* in this part, combustion takes place with a deposit of carbon. 4. A *conical envelope*, which gives but little light surrounding the whole flame, extremely thin or attenuated; combustion is complete in this part, and it is at its contact with the *luminous* envelope that the temperature is highest."

Another author says of the flame of a candle: "At its base we perceive a small part of a deep-blue color; in the middle is a dark part which contains the gas evolved from the wick, but which, not *being yet in contact with the air, cannot burn;* outside of this is the brilliant part of the flame. We also perceive on the confines of this latter a thin, faintly luminous envelope, which becomes larger towards the summit of the flame. It is there that the flame is hottest."

Dr. Thompson, in his work on Heat and Electricity, and other writers, give similar illustrations of the combustion of the gas in the flame of a candle; all of which point to an instructive lesson, much neglected in practice, and proves conclusively the necessity of intermingling air with inflammable vapors for the purpose of their combustion, as shown in the every-day occurrence in the flame of a tallow candle or common oil lamp.

It also proves that, although this flame be in contact externally with a current of air created by itself, yet a large portion of the tallow and oil passes off unconsumed, with great loss of heat and light. A similar waste is constantly going on in furnaces on a large scale.

All the authorities quoted agree on the main facts: First, that the dark center of the flame is unconsumed gas *ready for consumption*, and only waiting to get into contact with the oxygen of the air. Second, that the portion of the gas in which the due mixing has taken place, forms but a *thin film* on the *outside* of such unconsumed gas. Third, that the products of combustion form the transparent envelope (which may be perceived on close inspection). Fourth, that the collection of gas in the interior of the flame *cannot burn for want of oxygen*.

These points involve the whole case of the furnace, and illustrate the difference between imperfect and perfect combustion. It shows that, although the bodies of gas and air apparently have *free access* to each other, yet combustion is incomplete because *time is wanting* for their due mixture. Thus combustion proceeds according to the undeviating laws of nature, and only as the constituent atoms of gas get into contact with equivalent atoms of atmospheric oxygen.*

If, then, the unrestricted access of air to this small

---

* " In looking steadfastly at flame, the part where the combustible is volatized is seen, and it appears darker, contrasted with the part in which it *begins to burn*, that is, where it is *so mixed with air as to become explosive*." — SIR HUMPHREY DAVY.

flame by the laws of diffusion is not able to form a due mixture in time for ignition, it is clear that it cannot do so when the supply of air is restricted and that of the gas is increased.

In the Argand lamp, the air being partially under control and made to pass through a circular wick that is augmented, so that more gas is consumed *within a given space* than in the ordinary lamp; but why? Evidently because more opportunity for mixture is afforded through a series of jets or accessible points of contact. If the appertures through which air is admitted into the interior of the flame be closed, the appearance of the flame is changed — part of the supply of air being cut off, it extends further into the air before it meets with the oxygen necessary for its combustion.

Here we trace the imperfect combustion to the inadequate mixing — within a given time — of the *gas and air* until too late, or until the ascending current has carried them beyond the temperature required for chemical action; the carbonaceous constituent losing its gaseous quality, assumes its former color, and we have black smoke. Observing the means by which the gas is effectually consumed in the Argand lamp, it is evident that, by any device by means of which *gas jets* in a furnace could be presented to a sufficient quantity of *air* — as it is in the lamp — the result would be the same.

The difficulty of effecting such a distribution of gas by means of jets in a furnace is well-nigh impossible; but since the *gas* cannot be introduced to the *air*, may not *the air be introduced by jets into the gases.* How to do this successfully is the great problem to be solved in burning coal (especially soft coal) economically. This has been attempted, both in this country and Europe, with more or less success.

One of the most successful experimenters in this

6

direction, who has given the result of his experience to
the world, is Mr. C. W. Williams of England, to whom
the writer is indebted for much of the contents of this
paper.   In a letter to Mr. Williams, Prof. Brand, in commenting on his experiments, uses the following language : " Each jet of air which you admit becomes, as it
were, the source or center of a separate flame, and the
effect is exactly that of so many jets of inflammable or
coal gas ignited in the air; only in your furnace you
invert this ordinary state of things, and use a jet of air
thrown into an atmosphere of inflammable gas, thus
making an experiment upon a large and practical scale,
which I have often made on a small and theoretical scale,
in illustration of the inaccuracy of the common terms of
combustible and *supporter* of combustion, as ordinarily
applied.   In the one case (as in the Argand burner) the
gas in the center meets the air on the exterior; in the
other, the air in the center, issuing into the atmosphere
of gas, enlarges its own area for contact mechanically,
thus increasing its measure of combustion.   Thus we
see that the value of the jet, or the method of taking air
into the furnace through a large number of small apertures, arises in consequence of its *creating for itself* a
large surface for contact, by which a greater number of
elementary atoms of the *combustible* and " supporter "
gain access to each other in a given time.

Combustion, then, being a chemical process, depending
upon the mixture of the gases with air in the right proportion, and as we cannot control the gases in a furnace,
but can largely control the air : the question of perfect or
imperfect combustion, as far as human means can be
applied, is one regarding the *air*, rather than the combustible — the method in which it is introduced, as well
as the quantity and quality supplied — the control also
of atoms rather than of masses.

In this connection, it may be well to briefly allude to an erroneous impression that prevails in some quarters, that the hotter the fire — the more intense heat of the glowing coals — the more steam and power they get from a given amount of coal; also, that the gases are consumable by being brought into contact with a body of " glowing incandescent fuel." Yet any chemical work of authority would inform them of the well-established fact. that " decomposition, not combustion, is the result of a high temperature applied to the hydrogen gases — that no possible degree of heat can consume carbon — and that its combustion is merely produced by, and is in fact, its *union with oxygen.*

" It is a palpable over-sight of this distinction that has led to that manifest blunder — the supposing that the coal-gas in a furnace is to be *burned* by the act of bringing it into contact with bodies at a high temperature;" or in the words of the patentees, by causing it to pass *through, over and among* a body of hot glowing coals.

Indeed, those words of Watt, " through, over and among," have led many astray, occasioned much waste of money, loss of time, and misapplication of talent.

Such parties overlook the most important statements of Watt, in the same paper, viz., " *and by mixing it with fresh air,* when in these circumstances." The fact is, if the requisite air can be furnished and properly managed, the necessary heat for effecting combustion will never be wanting in the furnace.

SECTION II.

COMBUSTION OF COAL

PRACTICALLY CONSIDERED.

In the preceding chapter combustion has been treated mainly with reference to its chemical relations, and the necessary conditions requisite to burning the gases in bituminous coal. We shall now briefly consider its application *practically* in the construction of steam boilers and furnaces.

If we would economize fuel, we must give due attention not only to mechanical appliances, but also to the nature of the bodies we deal with, their constituent parts and chemical relations respectively; and as the laws of nature are inexorable, mechanical details must yield to those of chemistry.

With this principle in view it will be found that the furnace, in which the operations of combustion are to be conducted, is of the first importance. General improvements in the boiler and engines have been going on at a rapid rate, while scarcely any attention, until quite recently, has been given to the grates and the furnace. Even among engineers the notion too often prevails, that while every other appurtenance should be first-class, yet, " anything will do for a grate " that will stand up under hot fires.

Hence the attempt to economize fuel and power is confined almost entirely to the boiler and engine and their mechanical proportions, while the grate-bars and furnaces have been neglected, which latter are the real source of economy and power.

It is not our purpose here to go into the comparative merits of the different kinds of boilers or grates, but will briefly touch upon some fallacies or old notions, the value of which seems to consist mainly in their ancient origin. Some place their main reliance on *grate surface;* others on *large absorbing surfaces;* while a third demand, as the grand panacea, " *boiler room enough* " — without explaining what that means; and the requisites, in general terms, are summed up thus : —

1. Sufficient amount of internal heating surface.
2. Sufficiently roomy fire-box or furnace.
3. Sufficient air space between the bars.
4. Sufficient area in the tubes or flues.
5. Sufficiently large fire-bar surface.

All of which amounts to this : Give sufficient size to all the parts, and avoid deficiency in any.

" So gravely is this question of relative proportions insisted upon " — says a popular writer on this subject — " that we find many treatises on the use of coal, and the construction of boilers, laying down rules with mathematical precision, giving precise formulæ for their calculations; and even affecting to determine the working power of a steam engine, by mere reference to the size of the fire-grate and the internal areas and surfaces of the boiler; yet in this apparent search after certainty, omitting all inquiry respecting the processes or operations to be carried on within them."

One writer lays down the following dogmatic rules : " *For every cubic foot of water* to be evaporated *per hour*, allow one square foot of grate-bar; one square yard of heating surface; ten cubic feet of water space; five square feet of water surface; ten cubic feet of steam space."

Says another writer, commenting on the above : " Here we have all the proportions laid down and squared according to rule, as if it were the proportions of a building that were under consideration, rather than of vessels in which complicated chemical processes were to be conducted. These rules, however, will not teach us how best to effect the combustion of any given weight of fuel, or increase the generation, transmission, or absorption of any given quantity of heat. We have here laid down a scale of internal proportions, but no clue to that of the heat generative effect of a square foot of grate-bar, or the heat transmitting power of a square yard internal surface.

" It may, indeed, be asked, what relation a square foot of grate-bar can have to a cubic foot of water; or to any given weight of fuel? We know that under different circumstances, treble or quadruple the amount of these proportions may be beneficially or injuriously found in practice; and that even double the weight of fuel may be more advantageously consumed on a given area of grate-bars, in one class of boilers, than could be effected in another.

" In truth, the weight of fuel to be consumed has no legitimate relation to the space on which it may be laid, and depends on other considerations : viz., on the quantity of air passing through it, the time employed, and the weight of oxygen taken up by the several constituents of the fuel respectively.

" Again it may be asked, what relation a square yard of heating surface has to the transmission of any given quantity of heat, or the generation of any given quantity of steam? These calculations, in fact, have no value except on the assumed, but utterly erroneous data, that each square foot of grate surface is equivalent to the perfect combustion of a given weight of fuel, and the generation of a given quantity of heat in a given time; and that every square yard of internal surface must,

necessarily, be brought into action and received as equivalent to the transmission of a given quantity of heat.

"Now the magnitudes and quantities which here really
require to be calculated are *chemical*, not *mathematical*.
They are not those of flue-surfaces, or grate-bars, but
of the bodies to be introduced to them, the quantities in
which they respectively combine, and the heat evolved,
applied, or lost."

All boilers have their furnaces and grate-bars, on
which the fuel is placed; their flues or tubes through
which the flame or gaseous products pass; and the
chimney to obtain the necessary draught, and carry away
the surplus or refuse products.

The process of combustion is mainly carried on in the
furnace; hence the kind and quality of the grate-bars
upon which the fuel is to be burned, though oftenest
neglected, are of the utmost importance. Next, a roomy
ash-pit, to provide sufficient air and protect the bars.
Third, sufficient area of combustion chamber.

When all the conditions that belong to the *introduction
of air to the two distinct bodies to be consumed* (the gas
and coke) shall have been complied with, and systema-
tized in practice, there ought not to be any more difficulty
in securing perfect combustion in the *furnace*, than there
is now in the common gas burner. To solve this problem, however, instead of laying down inflexible rules of
proportions of the mechanical apparatus to be employed,
we must *first consider what is to be done with said apparatus.*

In the combustion of bituminous coal, we have seen,
there are two distinct bodies, the *solid* and the *gaseous*,
to be dealt with. With regard to the proportions of the
parts of the furnace, we have to *first* consider the *superficial area of the grate* for holding the *solid fuel;* second,
the size of *air spaces*, or proportional amount of space
to the iron; third, the means of keeping said air space

free from all obstructions to the draught; fourth, the *sectional area* of the chamber above the fuel, for burning the *gaseous portion of the coal and the introduction of oxygen to said chamber.*

As to the *area of the grate-bars*, as they are to support a *solid body*, no more area is required than the coal actually covers at a uniform depth, and it is alone important that it be *not too large;* while the area of the chamber above the coal, which is to be occupied by a constantly changing gaseous body, it is important that it be *not too small.*

With reference to the area of other parts of a boiler, no specific rules can be laid down with certainty, as the weight of fuel that may be consumed on any square foot of surface must depend on numerous other contingencies.

As to the size of the grate, observation of the engineer will enable him to determine the proper length or width of the furnace : the important consideration being that it be confined within such limits that it shall, *at all times, be well and evenly covered with fuel.   This is an essential and absolute condition of economy and efficiency;* yet in practice this condition is most often neglected.   If the grate-bars be not uniformly well covered, the air will enter irregularly and rapid streams or masses pass through uncovered parts, and at the very time and places when and where it should be most restricted.

Such a state of things bids defiance to all regulation or control of the supply of the *air, upon which we must depend entirely for perfect combustion and economy.*

The fire should not be allowed to run too low, as this would involve a loss of time and duty—the object being, of course, to obtain the greatest quantity of free-burning gases from a given amount of coal in a given time.

A charge of coal being thrown into a furnace, the heat from which the gas-generating process is effected is derived from *the remaining portion of the previous charge,* then in an incandescent state. This demand for heat, however, is confined to the commencement of the operation with each charge. The heat required for *continuous gasification* is, or ought to be, obtained chiefly from the flame itself; as seen in the case of a candle where the gasification of the tallow in the wick is derived from its own flame. Hence the importance of sustaining a suitable body of incandescent fuel on the grate, particularly when a fresh charge is to be added to the furnace.

We often hear complaints of the introduction of *air* being attended with decreased evaporation of water, or increased consumption of coal. These complainants should understand that in nearly all cases these results are due entirely to *inattention to the state of the furnace*, perhaps entirely due to the fireman leaving much of the *grate uncovered*, thus providing the shortest possible route for the introduction of *double* the volume that is required.

The uneven condition of the fuel on the grate-bars often causes great alterations of temperature, by permitting an *excess of air in irregular and uncontrollable quantities*, through the uncovered portions of the grate; when the fire begins to *burn in holes*, the evil increases itself by the accelerated rapidity with which the air enlarges the holes for its own admission, causing a still more rapid combustion of fuel around the uncovered parts, at the very time when these orifices should have been closed; hence the *importance* of keeping a uniform and even body of fuel on the bars, as well as the air spaces in the grate clean and free from ashes alike *all over the grate surface.*

With reference to the thickness of fires to produce the best results, opinions differ; still there can be no question that the fires should never be so thick as to prevent the *air from freely passing through the body of fuel*, and thus secure the proper mixture of the oxygen with the gases above the fuel, in order to obtain the most perfect combustion.

This leads to a brief consideration of certain alleged improvements in the way of mechanical devices for agitating or " shaking " the whole grate surface at once, and thus keeping the fire clean without opening the doors to " slice " fires. While all that is claimed for " shaking " or " rocking " grates may never be realized, still it is safe to conclude, that with all the improvements that have been made in steam boilers and engines, there have also been some improvements made in grate-bars. By the usual method of cleaning fires with the " slice " bar, there can be no doubt that much fuel is wasted and steam lost, as well as injury done to boilers, by the rush of cold air against the heated surfaces when the fire doors are open; hence any device, by means of which the above objections can be avoided, will not only pay on the score of economy, but greatly assist in the management and control of fires.

There are a number of movable or shaking grates in the market; but there are but two general principles or plans of construction: namely, the *horizontal reciprocal movement*, in which all the bars are moved at once in opposite direction each to the other, and lengthwise the furnace. The other a *rocking motion* of a series of bars or blocks running across the furnace. The first agitates or rakes the fire at the bottom only, sifting the ashes therefrom, and is *always level* whether in motion or at rest. The second has a semi-vertical motion, slightly opening and closing the air spaces. The latter must be handled with care — especially with fine fuel — or the upward motion will break up the fire too much and throw dust up the boiler tubes, and let good coal fall with the ashes. The best illustrations of these two general principles are the so-called " Roller Grate " and the " Banister Grate." All others are but copies or partial imitations of these two leading principles, and like most imitations, fall short of the original in merit.

In the light of what has been said on combustion, it will be seen that a practical " shaking " grate will supply some of the conditions requisite for getting the most *heat* and *power* from a given amount of coal, as by their use the fireman can carry a comparatively thin body of fuel distributed *evenly* over the grate surface, and free from holes. For by a slight movement of a lever — instead of the laborious effort with a " slice " bar — all the air spaces are cleared at once, without any waste of fuel, and a steady, even draught kept up, without loss of steam by opening the fire doors to " clean " fires. They also save injury to the boiler, caused by a rush of cold air to the heated surfaces while the fire-doors are open, thus causing sudden contraction, etc.

The uniform and continuous supply of fresh air through the grate bars to the glowing coals, with the least interference to the necessary conditions for a suitable mixture of the gases with the oxygen of the air, are the essential requisites for burning coal economically. A good " shaking " grate assists in providing these conditions. Besides the " shaking " grate, however, there is great need of some practical device for supplying air to the combustion-chamber above the coal — especially for bituminous or soft coal— to mix with the hydrogen gases and carbon which pass off unconsumed. This, added to a good shaking grate, combined with a good automatic damper regulator, will furnish conditions for combustion well-nigh complete and perfect.

In burning anthracite coal or coke, the principal points requiring attention are, the selection of grate-bars that furnish the best distribution of air, through small jets, to all parts of the furnace, and in sufficient quantity to effect a complete union with the gases, coupled with the necessary mechanical appliance for keeping them clean, so as to secure a uniform draught with the least interference to steady combustion and loss of steam and fuel. A properly constructed " shaking " grate will best secure these results.

Next to this, the good judgment of the fireman is most important, who, as a rule should carry a comparatively thin body of fire on the grate-bars, evenly distributed over their whole surface, so that the air can freely pass through the incandescent fuel and supply the gases with the proper amount of oxygen to effect their combustion. The fire should be kept level and free from holes. as the admission of large masses or streams of air has a cooling effect, and retards the generation and combustion of the gases.

The volume of air required for the combustion of the coke of a ton of coal (independently of the gas) may be easily determined, as in such a case there is but one combustible (the carbon) to be considered; and but one supporter (the oxygen) of the air. Therefore, any difficulty that may arise in practice is not a chemical one, but the result of some imperfection in mechanical appliances. While air is the main ingredient in combustion, it can be introduced in such a way as to quantity and method as to become a detriment, a fact often overlooked in practice.

One writer observes : " If chemistry did not teach us that the rate of combustion produced in the furnace is dependent on the *quantity of air passing through it*, every day's experience would soon convince us of this." Again, the same writer observes : " This being the case, the matter stands thus : the quantity of heat generated is dependent upon the quantity of air admitted. So also is the quantity of steam produced dependent upon the greater or less intensity of the fire." Another writer, commenting on the above, justly remarks: "Neither chemistry nor experience justifies these inferences. The quantity of heat generated is dependent on the relative weight of *hydrogen* first, and carbon afterwards, *chemically combined* with their equivalent weights of atmospheric *oxygen*. The quantity of air admitted may, indeed, actually diminish the quantity of heat generated. So the steam produced does not depend on the intensity of the fire, but on the quantity of heat absorbed;" and both chemistry and experience agree with this latter statement. Were there nothing to be considered in the use of coal but the combustion of the fixed carbon, nothing would be required but the supply of air through the grates to the fuel in proper quantity. In the use of coal, however, the gas, also, is to be generated and consumed, and any excess of air or its injudicious introduction, though it might not affect the combustion of the carbon, may materially interfere with the quantity required for the gas.

Now, as regards the quantity of air chemically required for the coke or fixed portion of the coal, after the gas has been expelled. It can be shown that every 6 pounds of carbon requires 16 pounds of oxygen. The volume of atmospheric air which contains 16 pounds of oxygen is estimated at 900 cubic feet, at ordinary temperature; and as bituminous coal contains about 80 per cent of carbon, we have 1.600 pounds of coke (the product of 2,000 pounds of coal) requiring its equivalent of oxygen, which will be equal to 240,000 cubic feet of air. This great quantity of air required for the exclusive use of the *coke on the bars* must be passed upwards from the ash pit, the product being transparent carbonic acid gas of a high temperature. The carbon remains quies-

cent and without combustion (irrespective of the temperature to which it may be raised) until each atom shall
successfully obtain contact and combine with its equivalent of oxygen, which becomes, as it were, the wings by
which it is literally carried away in the shape of carbonic
acid. Of itself, and without the aid of such wings, it
has no power of movement, escape or combustion.

The principal point requiring attention in supplying
the coke or solid carbon with air — as already noticed —
is preserving a uniform and sufficient body of fire on the
bars, so as to prevent the air passing through in masses
or streams, by which a cooling effect would be produced
injurious to the generation of gas.

The quantity of air required for burning the coke in a
ton of coal having been considered, we now enter a
more difficult field of inquiry, namely : to determine how
much air is required for the gas of the same. It has
been shown that for each cubic foot of gas, the oxygen of
ten cubic feet of atmospheric air is required. In the
process of making gas, it is understood that 10,000 cubic
feet of gas are produced from each ton of bituminous
coal, requiring no less than 100,000 cubic feet of air.
Adding this to 240,000 cubic feet required for the coke,
we have 340,000 cubic feet as the minimum quantity
required for the combustion of *each ton of coal*, besides
the excess which will always be found to pass beyond
the chemical requirement. It is not likely that this large
volume of air can, under any circumstances, all be introduced *through the fire bars* and fuel on the same. This
would not only be contrary to all chemical experience,
but involve a physical impossibility. It will be understood that a body of air can no more pass through a
mass of incandescent coke, without being deprived of a
large portion of its oxygen, than that the air can pass
through the lungs of a human being and yet retain the

necessary quantity of oxygen to sustain life in another. For this reason (as well as others) there should be some other channel provided for the introduction of air to the gases, as the impossibility of supplying the requisite quantity through the grate bars is shown for the most complete combustion of bituminous coal; and as the means of obtaining the largest quantity of heat from a given amount of coal, turns mainly on the *introduction of the air in proper quantity* and the mixing of the same with the gases in equivalent proportions, the manner of its introduction becomes of prime importance.

It has been stated (erroneously), that " the admission of air to a furnace should average from one-half to one square inch for each square foot of grate surface." Practice and experiment, however, prove that instead of *one square inch*, no less than from *four to six square inches for one square foot* of furnace is nearer the correct figures, the precise amount depending somewhat upon the gas-generating quality of the coal and the extent of the draught in each case.

The discrepancies in estimates made by experimenters as regards the amount of air per foot of grate surface, may be accounted for by a neglect in estimating the *velocity* of the heated gaseous matter passing through furnaces to the chimney. It is not the *egress* or escape of intensely-heated products we are considering, but the *ingress* or introduction of air at the average atmospheric temperature and pressure, subjected to impeded motion from friction in passing through small apertures. The following table of relative velocities of the air on entering will illustrate the joint influence of current and area of the air spaces : — *

---

*The quantity of air passing through well-constructed furnaces may, in general, be regarded as double what is rigorously necessary for combustion, and the proportion of carbonic acid generated, therefore, not one-half of what it would be were all the oxygen combined. — *Dr. Ure's Statement.*

Air aperture per sq. ft. of furnace for bituminous coal.	Velocity per *second* of ingress current of air at 60°.	Cubic feet per hour entering through small orifices.	For every ton of coal in cubic feet.
Square inches.	Feet per second.	Cubic feet.	Cubic feet.
6	5	7,500	75,000
6	10	15,000	150,000
6	20	30,000	300,000
5	5	6,250	62,500
5	10	12,500	125,000
5	20	25,000	250,000
4	5	5,000	50,000
4	10	11,000	100,000
4	20	20,000	200,000

Suppose a furnace measuring ten square feet of surface, with moderate draught, will be adequate to the combustion of two hundred weight of coal per hour, the gas from which would require 10,000 cubic feet of air. To supply that quantity within the hour will require the following relative areas of admission and velocity of current, viz. :

Velocity of current *per second*   Area of aperture in
of air entering the furnace.   square inches per
                                     foot of furnace.
If at 6.66 feet per second, will require 6 square inches.
" 10.   "   "   "   "   "   4   "   "
" 20.   "   "   "   "   "   2   "   "   .
" 40.   "   "   "   "   "   1   "   "

Thus the absolute necessity of ascertaining the practical rate of current of the air *when entering*, before we can decide on the necessary area for its admission. Then again, the *mode* of introducing the air must not be overlooked, whether it is introduced through *one* or numerous *apertures*,— whether in a *mass* or *divided form* — has a great deal to do with effecting complete combustion, and the utilizing all the gas of the coal. From any point of view, it is evident that the *manner* of introducing air to the furnace, as well as the *quantity* required, becomes an important consideration. The two bodies, the coke and the gases, are to be taken into account in the adoption of any mechanical appliance for introducing air or improving the draught and securing perfect combustion of all the elements of the coal. For bituminous coal, especially, suitable means must be adopted for effecting combustion of the gases, to secure the best results in the generation of steam and obtaining the most power from a given amount of coal with facility and economy.

Whatever the mechanical means may be, they should be arranged with the view of promoting a rapid diffusion or mixture of the air and gases. The irregularities in the generation of the gas in a furnace, and the constantly varying quantities to contend with, make it impossible to apply any inflexible rule which shall effect uniform results under all the varying conditions. The difficulties of regulating the admission of air by mechanical means, as to quantity, quality and mode for a suitable *mixture of the air and gases*, are obvious. The varying circumstances of land and marine boilers, of quick and slow combustion of large and small furnaces, the irregularities of the draught in different furnaces, and even in the same furnaces of the same boiler — considering, also, the various methods of firing and the uncertain qualities of the fuel used : all these and other unforeseen conditions render the theory of regulating the admission of air in accordance with mathematical calculations or rigid rules, mechanically applied, impracticable. Yet, such appliances are worthy of careful examination, and may be of great aid under the control of a competent engineer or fireman in the economical management of a plant.

It has been demonstrated by a great number of experiments, that to effect the perfect combustion of all the combustible gases produced in a furnace, a large demand for air (distinct from the air entering through the grate)

7

always exists. Also, that, by entirely excluding the air, smoke is produced and heat diminished in all states of the fire : hence, if correctly assigned proportions of air and gases and their due mixture are once ascertained, the attention on the part of the fireman is simplified, and far less required in regulating the admission of air. Among the many experiments that have been made and improvements devised for admitting the air to the gases through a hollow bridge-wall opening in the sides of the furnace, or perforated plates inserted in the boiler front or elsewhere, we know of none more successful than those of Charles W. Williams, brought out through patient study of the subject, and numerous experiments under the name of the " Argand Furnace," a number of years ago.

At the time, these experiments and devices attracted the attention of Sir Robert Kane (one of the highest chemical authorities of the day), who examined Mr. Williams' improvements and made an exhaustive report, which we here reproduce.

" The conclusions to which we have arrived, and which we believe to be established by very decisive evidence, as well of a practical as of a theoretical kind, may be briefly expressed as follows : —

" 1st. That in the combustion of coals a large quantity of gases and inflammable materials given out, which, in furnaces of the ordinary construction is in great measure lost for heating purposes, and gives rise to the great body of smoke which, in manufacturing towns, produces much inconvenience.

" 2d. That the proportion which the gaseous and volatile portion of the fuel bears to that which is fixed and capable of complete combustion on a common furnace grate, may be considered as one-fourth in the case of ordinary coal.

"3d. That the air for combustion of this gaseous combustible material cannot, with advantage, be introduced either through the interstices of the grate bars or the door by opening it. In the former case, the air is deprived of its oxygen by passing through the solid fuel, and then only helps to carry off the combustible gases, before they can be burned; and in the latter case, the air which would enter by reason of its proportionate mass would produce a cooling influence, and cannot conveniently be mixed so as properly to support the combustion of the gases.

"4th. That the combustion of the gaseous materials of the fuel is best accomplished by introducing, through a number of thin or small orifices, the necessary supply of air, so that it may *enter in a divided form and rapidly mix* with the heated gases in such proportion as to effect their complete combustion.

"5th. That in burning coke, or when coal has been burned down to a *clear red fire*, although the combustion on the grate may appear to be perfect, and little or no flame may be produced and no smoke whatever made, *there may be a great amount of useful heat lost*, owing to the formation of carbonic oxide, which, not finding a fresh supply of air at a proper place, necessarily passes off unburned.

"6th. That under the common arrangements of boiler furnaces, where there is intense combustion on the *fire grate* and but little in the *flues*, the difference of temperature in and around the various parts of the boiler are greater; and consequently the boiler is most subject to the results of unequal temperature. On the other hand, when the process of combustion is spread through the flues as well as over the fire grate, the temperature remains most uniform throughout, and the boiler and its settings must be least liable to injury.

"7th. That the heat produced by the combustion of the inflammable gases and vapors from the fuel in flues or chambers behind the bridge must be considerable, and can be advantageously applied to boilers, the length of which may be commensurate with that of the heated flues."

The writer, to further enforce these conclusions, goes on to describe the results of his experiments, made with boilers fitted up " with air-appertures (on Mr. Williams' plan)," namely, through the fire-doors, the boiler fronts, the bridge-wall and other devices for introducing air into the combustion chamber above the coal on the bars, in any suitable way to mix with and utilize the gases. These experiments are very instructive and were quite successful in accomplishing the end sought, namely : the burning of the gases, and the prevention of the smoke— in the latter even more successful, apparently, than any of the numerous "Smoke consumers " (?) of the present day, who, instead of giving due attention to the chemical principles and laws underlying combustion, which would aid them in devising means for *preventing smoke*, vainly imagine they can by mechanical means circumvent the laws of nature and *burn smoke* after it is produced.

In concluding this subject, we will draw attention to the importance of not only permitting an ample supply of air to the furnace through the grate surface or ash-pit, but also the necessity of some simple means — especially in burning bituminous coal — of allowing a large quantity of air to enter through the door or boiler front, by means of numerous small appertures (or by other effectual means), and thus secure the increased heating power arising from the combustion of not only the coke gas, but the carbureted hydrogen gases in the furnace.

From the many experiments and the experience of experts and authorities on this subject of combustion, it has been found, however, to be a matter of no special importance as regards effects, *in what part of the furnace or flues air is introduced, provided this all-important condition be attended to* — namely: that the mechanical *mixture of the air and gases* be *continuously effected* before the temperature of the carbon of the gas (then in the state of flame). *be reduced below that of ignition.* This temperature (according to Sir Humphrey Davy) should not be under 800° Fahr., since below that flame cannot be produced or sustained.

From all that has been said, we arrive at the following conclusions : First, that in the combustion of fuel there is but one body *combustible* to be dealt with, viz., the *carbon and hydrogen ;* and but one " supporter " required, the *oxygen* of the air. Second, that in combustion atmospheric air is the largest ingredient (yet it is the one to which, *practically*, the least attention is given, either as to quantity or control). Third, that both chemistry and experience teach that combustion depends, not so much on the quantity of air passing through the incandescent fuel, as on the *weight of oxygen which is taken up* in the passage. In fact, the quantity of air passing through it may be destructive of combustion if improperly introduced and distributed and when in excess of the demand of the fuel. Fourth, the quantity of heat generated depends, first, upon the relative weight of hydrogen and carbon ; afterwards, chemically considered, their equivalent weights of atmospheric oxygen ; so also, the quantity of steam produced does not depend so much on the intensity of the fire, as on the quantity of heat *absorbed by the water.* Finally, that success in generating the most heat and steam, and consequently power, from a given amount of coal, depends upon a compliance with the necessary conditions to perfect combustion, which involves not only a theoretical knowledge of chemical principles, but also a practical knowledge of the best methods of combining them with mechanical appliances, and the perfect mixing of the constituent elements with which we have to deal in strict accordance with the undeviating laws of nature.

PART III.

SECTION I.

THE THREE STATES OF WATER.

According to all chemical authorities, each *atom* of water is a compound of one equivalent of *hydrogen* and one of *oxygen;* and in dealing with its several states, we find it described — 1st. Crystalized (*ice*); 2d. Liquid (*water*); 3d. Gaseous or aeriform (*vapor*).

Physically considered, the properties of water when in a state of *ice* resemble those of other solid bodies in this respect, that its atoms are in close contact, have strong cohesive powers, and are incapable of motion. In the state of *liquid*, immobility is changed to mobility, with a strong attraction among its particles. In the state of *vapor*, other changes occur: attraction and mobility yield to mutual *repulsion* and *divergence*.

The presence of water (atoms) are found in almost all states and forms of matter by chemists, who sometimes find their presence a source of difficulty in their experiments. The quantities of heat inherent in water in each of its three states, in the general opinion of chemists, are: *latent heat* of ice 40°, of *liquid* 140°, and of *vapor* 1,000°. The first two are supposed to be correctly ascertained by physical tests; the last only by approximation, to what cannot be accurately determined. Heat is divided into two distinct classes, *latent* and *sensible:* the former signifying the *status* of water as a *liquid*, and producing no thermometrical effect; while the latter (sensible) exhibits its dynamic influence by its action on the thermometer or other bodies with which it is brought in contact if capable of conduction.

If then, the maximum heat in ice be 40° *latent* heat and 32° *sensible* heat, the inference would be, that each atom of the crystalized mass, on receiving an additional unit of heat, would have its statical condition changed; and losing its crystalized form, it would separate from the mass and become part of a fluid or liquid body. The same process would occur on its receiving a further unit of heat beyond what it is capable of retaining in the liquid state, and its status would then undergo a further change and become gas or *vapor*. In both cases—the passing from a solid to the liquid, and from that to the state of vapor—we notice the remarkable changes which supervene, exhibiting the peculiar characteristics of each.

Although the property in elastic fluids of "mutual repulsion"—the effect of which is termed *diffusion*—is generally recognized, it is also claimed that "*vaporized* bodies cannot be distinguished, on any scientific principles, from *permanent* elastic fluids." Whatever may be the cause of this principle of *mutual repulsion*, it is the main element which, in opposition to gravity, forms the chief characteristics of all *elastic fluids*, and should not be lost sight of, as it forms the *basis* of those effects exhibited in the various combinations of heat with liquids of all descriptions, from water to mercury.

Heat being applied to a body of ice, the cohesive property of the constituent particles *is lost*; the particles separating from each other, fall from the mass by the force of gravity and become a liquid. "Now, as the thermometric temperature of the ice was 32° and that of the liquid 32°, it follows that the entire amount of the heat communicated and by which the change was effected, must be considered in the latent state." If heat be continued, such of the liquid atoms as may receive each an additional unit will then assume the vaporous form; and as it had previously received its full complement of *latent* heat, this additional unit will consequently be *sensible* or available heat, and as such will act on the thermometer.

The *atom* of *ice* may be represented thus: One equivalent of hydrogen (H.), one of oxygen (O.) and one of caloric (C.); the *liquid atom*, one of H., one of O., two of C.; the *vapor atom*, one of H., one of O. and three of C., or the union of one atom of the liquid unit with three units of heat, two of *latent* and one of *sensible* heat.

SECTION II.

VAPORIZATION—WHAT IS VAPOR?

This subject has engaged the attention of philosophers and chemists of high authority; and yet there seems to be no general agreement of opinion respecting the theory of vaporization. The different writers have adopted such various methods of describing the process, and having made use of the term in connection with *evaporation* in a manner tending to confuse the mind and complicate the subject, so that much remains imperfectly understood respecting the effect of heat *on* and its connection *with* water.

Without undertaking to séttle the differences of the many able writers on this subject, we merely propose to present, in as brief and concise form as possible, the views of some of the leading writers, whose opinions have been accepted as authority, with the able review and criticism of Mr. Chas. W. Williams, who has treated the subjects herein presented in an elaborate and *original* manner, arriving however at entirely different conclusions in many important particulars.* This inquiry will embrace the following points : —

First.   What is vapor?

Second. How and where is it formed?

Third.   What are its special properties?

Fourth. In what does it differ physically and dynamically from water?

---

* "A treatise on Vapor Atoms, Heat, Water and Steam, embracing new views of vaporization, condensation, explosion, etc.," by CHARLES WYE WILLIAMS.

Fifth. What are the relative proportions of latent and sensible heat?

Sixth. What relation has vapor to electricity?

Strictly speaking, *vaporization* means the single process of converting atoms of a liquid into those of vapor. Numerous instances might be given of the misapplication of the term and the confounding it with others, especially with that of evaporation.

Had writers concurred in any one theory or definition of vaporization, there would have been less confusion. Turner* says: "Vaporization is conveniently studied under two heads — ebullition and evaporation. In the first, the production of vapor is so rapid that its escape gives rise to visible commotion in the liquid. In the second, it passes off quietly." Another writer (Mr. Williams) observes: "That vaporization cannot be studied under either head is evident, seeing that vapor may be formed without ebullition or any visible commotion whatever; and as to rapidity, — that being solely determined by the rate at which heat is absorbed by the liquid, — as much vapor will be generated in a given time by the same quantity of heat whether with or without ebullition. It may be broadly stated that neither ebullition nor evaporation have any immediate connection with vaporization." Dr. Lardner gives a different version of the subject, viz.: " When a liquid boils, vapor is formed in *every part* of its dimensions, and more particularly in those parts which are nearest the source of heat; but liquids generate vapor from *their surfaces* at all temperatures." How vapor can be generated at *the surface* of a liquid without a further application of heat is an unexplained mystery. Equally so when it is said, " Vapor is formed in *every part* of its dimensions." In such a case, where is the heat to come from by which

---

* "Elements of Chemistry," by EDWARD TURNER.

the liquid atoms are converted into vapor, or how is it
to arrive at the interior of a body of water?

In a popular work on Steam,* we have an epitome of the
almost universally received theory, which will serve as a
sample of all. " When heat is first applied to a body of
water, a rapid circulation of the fluid ensues. The water
at the bottom being first heated and expanded, becomes
lighter than the rest, rises to the top, and is replaced by
the current of cooler water descending to receive in its
turn a further accession. By and by small globules of
steam, formed at the bottom and surrounded by a film of
water, are observed adhering to the glass; as the heat
increases they enlarge; in a short time several of them
unite, form a bubble larger than the others, and detach-
ing themselves from the glass, rise upwards in the fluid.
But they never reach the surface; they encounter the
currents of water still comparatively cold, and, descending
to receive from the bottom their supply of heat, *shrivel up*
into their original bulk and are lost among the other
particles of water. In a short time the mass of the
water becomes uniformly heated; the bubbles becoming
larger and more frequent, are condensed with a loud
crackling noise; and at last, when the heat of the whole
mass reaches 212° the bubbles from the bottom rise *without condensation* through the water, swell and unite with
others as they rise, and burst out upon the air in a copious
volume of steam, of *the same heat as the water* from which
they are formed, and pushing aside the air, make room
for themselves," Mr. Williams, commenting on the
above, makes some pertinent remarks, as follows :—

1st. " Heating and expanding of the liquid are both
here assumed without proof or inquiry.

2d. " The water, becoming lighter, rises to the top and
is replaced by the colder water. No sufficient reason,

---

* "The Steam Engine," by JOHN SCOTT RUSSELL, F.R.S.E.

however, is given for this replacement. An ascending lighter body would necessarily remain at the top, as a cork would.

3d. "Globules of steam never adhere to anything: they have no such power or property. It is only when reconverted into the liquid state that adhesion becomes available.

4th. "Globules, either of water or air, remain always visible up to the surface.

5th. "The idea of bubbles of steam being condensed in their ascent is wholly inadmissible and contrary to fact.

6th. "As to the steam being of the same heat as the water from which they are formed: that is simply impossible, unless by ignoring the effect of heat."

This same writer goes on at length with the analysis and gives a more reasonable view of the process as follows: "Water, undistilled and unfiltered, being put into a glass beaker, over an Argand burner, numerous small globules will shortly be seen adhering to the bottom and sides of the glass. These have been mistaken by many writers for new-formed vapor, and, as above stated, for *globules* of *steam*. They are, however, mere globules of *air*, invisible at first by reason of their minuteness, but becoming enlarged as the glass to which they adhere becomes heated, and further, increasing by accumulation, they become visible, and adhering to the glass with such tenacity (if the process be carried on gently) that they may even be touched with a fine wire and swayed from side to side before they are dislodged. These not unfrequently remain adhering to the bottom until ebullition has begun to agitate the mass. That these globules have no relation to vapor is proved by the fact, that if, by being previously boiled and filtered, the water has been deprived of its air (of which it contains about two per cent), and if on being cooled the process be repeated, no globules will appear."

It has already been stated that atoms of liquid becoming atoms of vapor by the addition of heat, their characteristics are entirely altered; mutual attraction and mobility being changed to mutual repulsion and separation, with an increase of volume to an extent which makes them lighter than the surrounding atoms of liquid.

On these newly acquired properties depend the whole phenomena which steam exhibits. " Steam," according to Prof. Dalton, being " An elastic fluid like common air, and possessed of similar mechanical properties."

On this Sir Robert Kane observes, " The particles of volatile bodies repel each other at all temperatures, *until they occupy completely the space in which the body is contained,* and exercise a pressure which is equal to the force of their mutual repulsion, and which is termed the *elasticity of* vapor." We here recognize the elements of divergence or diffusion, force and pressure in volatile bodies.

An important question then arises, namely: whether atoms of vapor on their formation retain and exercise their several properties *as an elastic fluid,* while they remain *in a body of water* in which they have been generated, before and until their escape into the air. It would appear from the foregoing that a rigid inquiry into the process of the union of heat with liquids is necessary.

As the change from the liquid to the vaporous state is the direct result of the union with further increments of heat, it is a matter of indifference from whence that heat may be derived, whether *from above,* as from the rays of the sun or temperature of the air, or *from beneath,* as when heat is artificially applied.

When heat is applied from above, it is evident the upper or surface liquid atoms must be in *absolute contact* with the air which rests upon it; the heat radiating downward upon the atoms forming this *surface,* each will absorb one or more units of heat, converting it into a state of vapor with its properties of increased volume and levity intact.

This is a clear case of vaporization. The atoms of liquid being converted into atoms of vapor, and being subject alone to the weight of the atmosphere, there is nothing to prevent the full development of their volume.

The enlarged volume, arising from the difference between the states of liquid and vapor, has been estimated as the difference between a cubic inch and a cubic foot; or the bulk increased 1,728 times. Whether this estimate as to the enlarged volume be reliable or not, it is clear that in the vaporization of this surface stratum being effected its atoms will rise into the air and be replaced by the next in succession until the whole has passed away into vapor. In this way, lakes or pools of water are sometimes vaporized, the ground dried and the atmosphere replenished with vapor, which in turn descends in the form of dew or rain.

We next examine the process when heat is artificially applied *to the bottom* of a vessel containing water. Here the liquid atoms forming the lowest stratum are spread upon the bottom (like a carpet) and nearest the source of heat; the absorption of the heat, the changes in the character and form of the liquid atoms, is the same as when applied to the upper stratum; except in the latter case each atom receives its heat *direct* from the source *above it*, while in the former case each liquid atom receives heat *by conduction* through the vessel containing the water. Special attention however, is directed to the different *conditions*, as they involve the main feature of the theory here contended for.

The *surface* stratum of liquid atoms on being vaporized rise in the air as a bird from the ground, or a balloon on obtaining the requisite levity, alone impeded by the surrounding pressure of the atmosphere; while the *lowest*, or carpet stratum, have a new or different element and obstruction to contend with. They are not in contact with the *light medium of air* and a pressure of fifteen pounds to the square inch, but in a *medium* of *water* which has a density eight hundred and thirty times greater than that of air. The result is that an atom of vapor, generated at the bottom of a mass of water, has to force its way upwards through this dense medium to the surface before it can come in contact with the air.

While the surface atoms on becoming vapor were enabled to expand, say 1,728 times that of their liquid volumes under atmospheric pressure, it is evident that those formed at *the bottom* must be influenced by the *additional density and pressure of the liquid* medium in which they are generated and through which they have to work their way. Besides, they not only had to ascend, but diverge and diffuse themselves by virtue of their mutually repellent principle. So numerous, however, are these vapor atoms and so rapid is their generation, that a portion of them are found reaching the surface and escaping into the air almost instantaneously after the heat has been applied. This has been fully demonstrated by experiment* in which we have visible and physical proof : —

1st. Of the rapidity with which vapor is formed.

2d. Of its diffusion and consequent homogeneous temperature throughout.

3d. Of the identity in a dynamic point of view as *an elastic fluid* of the vapor thus formed and escaping after having passed through the liquid mass in each vessel, and also,

4th. That it is *vapor* and not *heated water* that is seen rising through the water.

---

* "Water and Steam : Vaporization, Evaporation, etc.," by C. W. WILLIAMS, A. I. C. E.

SECTION III.

VAPOR: AN ELASTIC FLUID.

The vapor of water all authorities admit to be an *elastic fluid*, with which is associated *pressure* or *force*. The term *elasticity*, however, has no legitimate reference to such properties in connection with a *single* body, or with liquids or vapor *en masse*. The term elasticity is correctly defined as "the force in bodies by which they endeavor to restore themselves to their previous position or form." When speaking of a sponge or a spring, this is intelligible; but air and vapor are different bodies and must be considered as distinct substances. Both are an aggregation of separate bodies, each of which is endowed with a property of *repulsion* among the atoms of its kind, through which it becomes the element of what is termed elasticity of the body. This should be borne in mind, as we direct the attention to the investigation of liquid, gaseous or aeriform masses.

On this point Prof. Farraday observes: " We have but very imperfect notions of the real and intimate conditions of the particles of a body existing in the solid, the liquid or gaseous states; but when we speak of the gaseous state as being due to the *mutual repulsion of the particles* or of their atmospheres, although we may err in imagining each particle to be a little nucleus to an atmosphere of heat, or electricity, or any other agent, we are still not likely to be in error in considering the elasticity as dependent on *mutuality of action*."

This bears directly on the question of *unity*, the term used in representing *heat* in vapor, and the *mutuality* of action in such units in the aggregate or mass.

8

Prof. Rankine says : "A perfect gas is a substance in
such a condition that the total pressure exerted by any
number of portions of it at a given temperature against
the sides of a vessel in which they are enclosed, is *the
sum of the pressure which each such portion would exert* if
enclosed in the vessel separately at the same temperature." Dalton's theory in reference to fluids of any
kind : " ' Each such portion ' must have reference to *each
separate atom* or particle." It is the same thing as to say,
that " the pressure exerted by any gas or elastic fluid is
the sum of the repulsive forces which the several atoms
exercise when *en masse*." As each atom of vapor, then,
represents *a unit* of heat and repulsive force, so the
amount of heat or pressure must be the *sum* of the units
exercising such force. Temperature and pressure, therefore, are but co-efficients of the *quantity and number of
atoms* present in any given space. The Professor's
legitimate inference is, that "divergence or expansion
is a property independent of the pressure of *other masses
within the same space.*" This is also in accordance with
Dalton's law, "that each gas or elastic fluid *enters as
into a vacuum.*"

Now the practical results of vapor is involved in this
law governing gases and elastic fluids, and lead to the
following questions : —

1st. Is vapor an elastic fluid?

2d. Has it the properties of other elastic fluids?

3d. Does it exert those properties independent of
other masses in bodies in the same space?

Therefore the analogy between vapor and other elastic
fluids should be ascertained.

" *The density of the air,*" according to the latest
authority, " is the result of the pressure to which it is
subject." The air is an elastic fluid; that is, its bulk

increases and its density diminishes whenever the external pressure is wholly or partially removed. — *English Cyclopedia :* " Air."

Again : " The *repulsive force* of the particles of air, of which we know nothing but its effects, is a counter-balancing force *from within* to the *pressure from without,* and is greater or less according to the greater or less *nearness* of *the particles.* In other words, the elastic force of the atmosphere as distinguished from the superincumbent column of air." *So of the vapor of water.* As an elastic fluid, there is the *repulsive force* of its several atoms acting as a counter-balancing force *from within* to the pressure *from without.* We may then consider this mutually repellent action as being the direct source of what is called the pressure of the mass, and that *vapor* but follows the same general law of other elastic fluids when relieved from pressure *from without* (that is, the surrounding medium, whatever it may be), its bulk increasing and its density diminishing as the external pressure is partially or wholly removed. This is precisely the case when the vapor particles, on their escape into the lighter medium of air, are removed from the denser medium of the water.

The same authority adds : "As we ascend in the atmosphere, the superincumbent column of air becomes of less weight and the density becomes less; that is, a cubic foot at the height of 1,000 feet above the ground is not so heavy nor *does it contain so much air* as a cubic foot at the surface of the earth." The same of vapor, as it rises from the bottom of the vessel (in which it is generated) to the surface and thence into the air. If the air presses equally on all sides and in all directions, why may not the same be said of the denser medium of the water acting on the several particles of vapor in it? The question of bulk then, is a question of quantity or number in given spaces, each particle, however, preserv-

ing its identity from the moment of its generation until
it reaches the highest region of the air. Again: "The
air having in itself a force which *tends* to *separate its
particles* from one another, or to *expand the whole bulk*,
but which grows less and less as the particles are more
and more separated, that is, as the bulk increases." Now
what is this *force*, which so tends to separate its particles
from one another, but the *mutually repellent property*
inherent in the constitution of the vapor atoms? As we
know of no power in nature capable of producing this
tendency of particles *to separate or repel each other* but
electricity, we have but to substitute the elastic fluid
vapor for that of *air* and the description is reasonable
and the analogy is complete.

When we speak of a body of water, filled or saturated
with vapor, we may equally describe it as we do air,
namely: as being that state in which " the elastic force
on a square inch of the surface of the air arising *from
its own constitution* just balances the pressure upon that
square inch." In other words, as the state of equilibrium
which just balances the pressure of the medium in which
it exists, whether that medium be water or air.

It is said that from careful experiments, it appears
that air and all other gases, as well as vapors, and all
mixtures of gases and vapors, obtain an increase of
elastic force for every increase of temperature, and expand, if possible, in the vessel that contains them.

Throughout we find that *increase of temperature* is regarded as the basis or cause of the several changes of
pressure, expansion or elastic force. The imparting of
heat, however, is one thing; but the indicated temperature or amount of heat, quite another. Temperature as
shown by the thermometer is but the index marking the
several changes as they are produced. Dalton remarks,
" It appears to me as completely demonstrated as any
physical principle, that, whenever any two or more gases

or vapors are put together into limited space, they will
finally be arranged each as if it occupied the whole space,
and the others were not present." *Now the strict application of this law is what is contended for here.* If vapor
be an elastic fluid, and endowed with all the properties
common to its kind, why shall not this law be equally
applicable to the vapor of water as to any other known
vapor, each being "arranged as if it occupied the whole
space and the others were not present." The whole
question turns on this : whether the vapor of water acts
the part of an independent gas, and retains its properties
while *in a medium of water*, as others do. That it does so
maintain its identity and individuality, can be as "completely demonstrated as any physical principle can be."
This may be considered self-evident from the re-appear-
ance of the vapor itself rising, with all its properties, on
being liberated from the water; for we can make no
distinction between the vapor arising from *before* or *after*
the heat has been withdrawn. That gravity has nothing
to do with the mixing and diffusion of gases or vapors in
the medium into which they are introduced, was proved
by Dalton in his reply to Priestly.

It is stated that Prof. Graham investigated the phenomena of diffusion with extreme precision and determined
that the diffusive volumes are inversely as the squares
of the densities of the gases. Dalton says, "If a quantity of water freed from air be agitated with any kind of
gases *not chemically uniting with water*, it will absorb its
bulk of the gas." Why, then, shall not the same reasoning and the same law be applied to the mixing and agitating of the *elastic fluid*, *vapor*, with water : why not say
that, if a quantity of *vapor* be agitated with water, it
will absorb or take up the bulk of the vapor. Now, in
spite of opposite theories and prevailing opinons, this
mixing by agitation of water and vapor will be found
literally and strictly true.

SECTION IV.

ON HEAT AND EXPANSION OF WATER.

The prevailing theory on this subject is: that water, while *still retaining its liquid form and character*, absorbs heat and expands in proportion to the quantity of heat absorbed up to the temperature of 212° — the amount of the expansion being equal to $\frac{1}{2250}$, or according to Dr. Ure, $\frac{1}{25}$ of its value. There are however, so many proofs that may be adduced in contradiction of this theory, that it may be considered at least an open question and will bear further examination.

1st. With reference to the properties of *compressibility* and *incompressibility*. It is generally admitted that *water* is so little susceptible to compression that it may be considered practically incompressible. But as nearly all writers, while admitting the incompressibility of water, still insist on its expansibility, it may be well to refer to some of the recognized authorities and examine the grounds on which this theory is founded.

It has been found by experiment that under a pressure of 2,000 atmospheres there was scarcely an appreciable amount of compression in water; and what little there was may be reasonably attributed to the portion of air and vapor it contained. Dr. Lardner observes: "All *solid* bodies being gradually heated from the temperature of melting ice (32°) to that of boiling water and then gradually cooled down from 212° to 32°, will be found to have exactly the same dimensions at the same temperature during the process of heating and cooling, the gradual diminution of bulk in cooling corresponding exactly with the gradual increase of bulk in heating. Glass and other bodies, gradually heated from 32° to 212°, which undergo degrees of expansion of the solid corresponding to *two* degrees of the thermometer, is *twice* the expansion which corresponds to *one* degree, and so on, the quantity of expansion being multiplied in the same proportion to the degrees through which the temperature had risen is multiplied." This rule applied to liquid water is strictly analagous, as follows: The number of its atoms converted into vapor corresponding to *two* degrees of the thermometer is *twice* the number (and twice the volume) that corresponds with *one* degree, the number vaporized being multiplied in the same proportion to the number of degrees through which the thermometer has risen. Again, he observes: "The force with which a solid dilates is equal to that with which it would resist compression; and the force with which it contracts is equal to that with which it would resist expansion." This is simply action and reaction expressed in law, dilatation and compression being correlative terms.

Now this correspondence in *forces* being a general law of nature, must be applicable to all bodies; and what are the elementary constituents of liquid but bodies subject to the same law?

Here we see how a positive law in physics may be rendered negative or doubtful when applied to liquids, when an arbitrary application is resorted to in order to satisfy the *theory* of expansion. It is here inferred that the resistance to expansion *as regards liquids* is not commensurate to that of compression; or, in other words, that action and reaction are (in this case) not equal and opposite. If these liquids be incompressible

they must be inexpansible; for if the resistance to compression cannot be overcome, neither can the resistance
to expansion or dilatation be overcome. This idea
would be more in harmony with nature's laws.*

In his paper on EXPANSION, it is shown that all writers
concur in saying that vapor and air follow the same law
as gases or vapors of all kinds. Pursuing this inquiry,
Prof. Thomson remarks, that "Dalton and Lussac, by
keeping the gases experimented on dry, were enabled to
discover that all gases experienced the same augmenta-
tion of bulk when subjected to the same temperature.

---

* In testing this recognized theory further with reference to
CONDUCTIBILITY and NON-CONDUCTIBILITY, Mr. Chas. W. Williams,
in his book on this subject, calls in question the recognized authorities; and after quoting extensively from such writers as Prof.
Brand, Sir Robert Kane, Dr. Read, Dr. Ure, Prof. Graham and
others, showing the harmony and uniform agreement in their theories and "erroneous assumptions," — in regard to the "ascending
and descending currents" which claim the "circulating currents"
are the cause of the "uniform temperature of the mass," etc., —
goes into the subject very thoroughly and exhaustively, showing
by logical argument and numerous experiments and illustration
that the whole theory is an error, — unscientific and misleading, —
confessing withal that he formerly readily adopted the same
theories, being misled by the authoritative statements of previous
writers on the subject. Mr. Williams finally sums up as follows:
"That ascending and descending currents do ultimately prevail is
certainly true; equally so that they are mainly instrumental in
effecting circulation, in which sense the theory was adopted."
He adds: "Subsequent experience, however, distinctly proved
that they have no reference to that homogeneity of temperature
which prevails long before these circulating currents begin, and
which are solely attributable to the action of ebullition." He further
says: "In all these statements we see everything is assumed: the
heating and expanding of the water, the ascending and descending
currents, and the supposed contact of every particle of the water
with the bottom of the vessel." The value of these experiments,
as Mr. Williams himself remarks, "consists not so much in dis-
posing of the theory of *descending currents*, as that it furnishes
conclusive evidence both of the existence of the *vapor* in the
water and its diffusive action throughout the mass."

Hoping to find a similar coincidence in liquids, the subject was pursued with great labor, forgetting, however, that the atoms of gases, vapors or other elastic fluids have no *fourth state* into which they may be resolved by additional temperature; whereas *liquid atoms*, by heat alone, become virtually atoms of an entirely different class of bodies, and possessed of essentially different properties — as different in fact as are the elements of water, oxygen and hydrogen, in their separate states as gases, and their combined states as liquid water."

Baffled in the attempt, Thomson comes to the conclusion that "liquids differ from gases in this, that their expansibility is *not uniform*, but that the *rate of expansion* increases with the temperature and is, therefore, the greater the higher the elevation at which a given quantity of heat is added to them." He saw the difficulty of reducing the *rate* of expansion in liquids to the law which regulates that of gases. Yet by applying the law of the *quantity* of *vapor* present in any given body of a liquid, a sufficient solution of the rate of expansion would have arisen. This doubtless hereafter will be determined.

Dr. Lardner remarks: "The same vessel will hold a greater quantity of cold than hot water. If a kettle filled with cold water be placed on the fire, the water, when it begins to warm, will *swell* and flow from the spout until it ceases to expand." If by expansion the enlargement of the gross bulk of the water only is implied, this is true enough. But such expansion is the result of elementary atoms being successively converted into vapor. In no other sense does the water *swell*. And in *that* sense it never ceases to swell all the time heat is applied until the point of saturation is reached.

Again he says: "Since the magnitude of any body changes with the heat to which it is exposed; and since when subject to the same calorific influence, these dilatations and contractions, which are the constant effect of heat, may be taken as a measure of the physical cause which produces them . . ." This is doubtless correct when applied to individual bodies among which the *constituent particles* of liquids may be classed, but not to the *aggregate of* those bodies. Here lies the main source of error. If water were a body to be dealt with *in bulk*, as a ball of iron or lead, and capable of receiving and *conducting from* atom to atom successive increments of heat, without any change in its *status of liquidity*, we might, in such a case, correctly infer its expansion. But water, or indeed any liquid, has not that power or property of conduction among its constituent particles as metals or solids have, and consequently is incapable of expansion in the sense of such bodies. Besides, heat, that invisible and imponderable agent, knows nothing of the mass of contents in the vessel. It deals only with the *individual atoms* of which the mass is composed, whether liquid or solid, and with which it comes into contact. When, also, we consider how nature in its wonderful economy apportions the combining volumes, weights or other properties of matter, there can be no disproportion between atoms of liquids or solids, and units of heat. Their union is but part of the immutable law of nature stamped on matter of all kinds. Each atom has not only its specific duty to perform, but the faculty of performing that duty; none will be tried and found wanting. The power of the wind is but the *sum* of the powers inherent in each *individual atom.* So of the waves, or a crowd of human beings. It is to these, then, and their respective properties, to which our inquiries should be directed.

As water or its constituent particles cannot undergo any change, physical or dynamical, without a sufficient cause, liquid particles at the temperature of 32° must continue at 32° until they have each received their equivalents of heat by which they lose their *status* of *liquidity*. They are then no longer liquid atoms, but absolute *atoms of vapor*. On what grounds, then, can we assume that *liquid atoms* are heated or expanded, and still retain their liquid form and properties? Such an hypothesis would be contrary to the evidence of facts.

To say that water can be a recipient of heat and be expanded while it retains the liquid state, and is also a *non-conductor of heat*, would involve a physical solecism irreconcilable with reason and common sense.

The oversight of the various authors who have written on this subject consists mainly : —

1st. In ignoring the formation of vapor as rapidly as heat is applied.

2d. In overlooking the existence of *vapor atoms* in the body of the water, and their diffusion throughout the mass.

3d. In assuming the liquid atoms to be the recipient of the heat without any change in their *status*.

4th. In overlooking that *vapor atoms*, being necessarily individually larger than the *liquid atoms* from which they were formed, fully accounts for the gross enlargement of the mass.

The writer, from whom we have made copious drafts, in treating this subject (and who is chiefly responsible for the views herein) has finally summed up the argument in the following manner : "Correctly speaking, then, there can be no such thing as heated or expanded water or other liquids, even as regards mercury, which only follows the general law.

In ordinary language and as a mere conventional term, we may speak of water being heated and expanded, and of having an increased temperature. When, however, we are treating the subject in a scientific point of view and with reference to the strict nomenclature adopted by chemists, we should avoid whatever may be unwarranted as but tending to confusion, as this may lead to serious complications. Water may be spoken of as being mixed *with vapor* as *with air*, and its variations of temperature described. That temperature, however, should be attributed to its right source, namely: the quantity or number of vapor atoms then present in any given space, these being the true and only source of dynamic effect.

On the whole, then, we have sufficient to justify the following conclusions : —

1st.   That water, or other liquids, being incapable of *compression*, are equally incapable of *expansion*.

2d.   That water being a *non-conductor* of heat, must also be a *non-recipient* of it.

3d.   That as it cannot be heated or expanded and still retain its liquid form and properties, it cannot be thermometrically affected.

4th.   That its enlarged volume is attributable, not to any measure of expansion *as a liquid*, but to the presence of *vapor* in it, in a state of an *elastic fluid*.

5th.   That this condition is in entire accordance with the recognized laws of elastic fluids.

The respective properties of liquid and vaporous atoms, as regards the changes they undergo by heat, from the liquid to the vaporous state, may be thus described : —

LIQUID ATOMS.	VAPOR ATOMS.
1. Gravity.	1. Gravity.
2. Latent Heat.	2. Latent and Sensible Heat.
3. Mutual Attraction.	3. Enlarged Volume.
4. Mobility, *inter se.*	4. Increased Temperature.
5. Non-Conductibility.	5. Mutual Repulsion.
6. Incompressibility.	6. Diffusion or Divergence.
7. Inexpansibility.	7. Conductibility.
8. Negative Electricity.	8. Compressibility.
	9. Expansibility.
	10. Positive Electricity.

SECTION V.

HEAT UNITS; VAPOR ATOMS.

More than a century ago Dr. Black announced his discovery in heat, and the laws have ever since been the subject of investigation and experiment. Since then, numerous writers have thrown more or less light on the phenomena of heat; " but the *relations between the pressure, the density,* the heat and the temperature, are yet undetermined and but very imperfectly understood." The common, but erroneous, theory that water becomes the recipient of heat, while still retaining its liquid form, and the assumption that the relations between the pressure, the density, heat and temperature are influenced by separate laws, are not calculated to throw much light on the subject.

One writer says: " Water *retains its heat only under pressure.* If the pressure be relieved, the heat, which it is then unable to retain, is carried off by the formation of steam, until the increasing pressure and the decreasing heat are again in equilibrium." Another writer (Mr. Williams) very appropriately remarks: " With equal regard to fact might it be said that a pound weight of *shot* retains its heat *only under pressure.* In truth, pressure has no more relation to the chemical absorption or retention of heat by the atoms of water than by the grains of a body of shot. This theory appears based on the double mistake: first, of assuming water to be a *body* in the sense that we speak of a body of lead, rather than as an aggregate of separate bodies, say of a given weight of shot; and secondly, that it is only when it can *retain no more* that the heat goes to the formation of steam."

Again it is said, " That pressure is maintained on
water *by its own steam.*" Were this the case, how are we
to account for the formation and visible appearance of
steam, from *the moment the heat is applied*, and while the
temperature is raised but a single degree from the
starting point?

Now the uniformity of temperature that prevails *in
the water* and *above it*, in close vessels, ought to be conclusive as to a corresponding pressure, or, what is the
same thing, a correspondence in the *quantity* of *vapor* in
both places, in excess of saturation. If this were not so,
Dalton's law of the " water acting the part of a vacuum
to elastic fluids " would be erroneous.

A certain number of degrees, 212, is said to indicate
the maximum temperature of water under atmospheric
pressure, but in exceptional cases it has been shown a
much higher temperature may be reached *without ebullition*
or any increase of pressure. Dalton correctly observes :
" All bodies are constituted of a vast number of extremely
small particles, or atoms, bound together by a force of
attraction. Besides this we find a force of *repulsion*.
This is now generally, and I think properly, ascribed to
the agency of heat." So that we have to consider the
matter of water *not in the mass*, or as a single body; but
as an aggregate of bodies, the constituents of which and
their several relations to heat are the proper objects of
inquiry.

Therefore, we must consider the action of *the smallest
elementary portions of the one with the smallest portions of
the other;* and the inquiry into the constitution of .the
matter of water, with reference to heat and temperature,
should relate to the molecules *or atoms*, rather than to
the mass. With these as *independent bodies* the heat
has combined, and the result of such union is not that the
whole body has been heated, but that such heat is con-

tined to the *individual particles* affected and converted into
vapor. We must not lose sight of the fact that a body
of water or any liquid is as much, so far as practical
results are concerned, an aggregate of individuals as
an army or a multitude of people.

We are not inquiring into the nature of heat, but into
the effect physically and dynamically produced by its absorption or union with the atoms of which water is composed; therefore, in this case, as with all other descriptions of matter, " *equivalent determinate quantities* are
essential in producing given results." We have no reason
to doubt that the same law applies to heat, in its combinations with the matter of water. If we know the
gaseous constituents of the elementary atoms of water,
there can be no rational grounds for objecting to treat
such atoms in their respective unions with heat.

It may be said that, as heat is an imponderable element,
we cannot speak of it as " atoms of heat;" yet, in a
scientific inquiry into the relative quantities absorbed or
brought into union with equivalent quantities of water,
we are justified in speaking of it as in *divided portions*.
This is the common method of all writers on the subject,
under the term " units of heat" (sometimes termed
doses*), as so many different increments of temperature.
By this system of equivalent atoms in all combinations of
matter we are enabled to decide chemically and physically,
on atomic processes, and to understand how and why the
air we breathe, and by which life is sustained, in its elements is identical with that destructive compound, *nitric
acid*, by which animal life would be instantly destroyed;
and that the only difference between these life-sustaining

---

* " Most bodies are susceptible of three states of existence ;
namely, solid, the liquid, and the elastic, or vapors ; and all these
are effected by the introduction of different *doses of caloric*."
—*Rees. Cyc. Condensation.*

and life-destroying compounds is one of *mere definite proportions* or equivalents. Therefore we are justified in speaking of one or more *units of heat* in combination with one or more *atoms of vapor*; and to understand the properties of water in any of its states, in combination with heat, we must consider the mutual action of *units of heat* in the constituent particles or *atoms of water*.

Heat acts on and vaporizes the liquid atoms in the human frame in the same way and under the same immutable law as it does in liquids, in any form or phase.

When the thermometer indicates 212° in liquids (as in the blood in the animal economy at 98°) further increments of heat are absolutely available. not as necessarily influencing or increasing temperature in the mass (which they could only do by remaining in it), but as generating further atoms of vapor : these, however, being no longer retainable by the laws of diffusion, and under mere atmospheric pressure, rise, and escape as rapidly as they are formed, each carrying away its respective equivalent of heat.   It is, then, not the quantity of heat that is influenced by the pressure, but the quantity. or number of atoms of vapor in the mass which influences that pressure.   Each individual atom of a liquid. when brought into union with an individual unit of heat, may be considered as a *distinct* entirety, or substance, and no doubt bears a given relation to temperature, although imperceptible by our powers of vision or measurement.

These facts are well understood by chemists; yet nearly all continue to regard liquids as separate bodies or integers, rather than as aggregates.   These generalities must be abandoned, and we must look at water in its elementary particles as we would at those of other descriptions of matter.   The necessity for this mode of proceeding will be more apparent when we consider that the quantities communicated are divisible into distinct classes, namely: *latent* and *sensible*, or more correctly speaking, *statical* and *dynamical*, the former being identified with the *status* of water as a liquid, and producing no thermometrical effect whatever; while the latter, however limited may be the quantity, has its dynamic influence shown by its action on the thermometer, or other bodies (if capable of conduction) with which it may be brought in contact.*

The foregoing analysis seems to warrant the conclusion that even the smallest quantity of water may, as a series of individual atoms, be vaporized and dissipated by proportional equivalents of heat, by which we are led to recognize the conversion of *single liquid atoms into vaporizations* as the only means by which each can obtain the buoyancy that enables it to rise and pass into the atmosphere. What takes place in vaporizing a single atom must be applicable in the case of millions constituting a body or a single drop of water, and is but a type, illustrating the principle or process going on with water at the boiling point.

The requisite equivalents of latent and sensible heat being absorbed by, associated with, or united to the liquid atom, its conversion into one of vapor *is complete;* and from the state of an *inelastic* body, with the property of *attraction* and *mobility*, it has assumed that of an *elastic* fluid with the opposite properties of *repulsion* and *divergence*. From this point of view, we are enabled to appreciate and apply Dalton's rule, that "*the force and pressure of steam is the same in equal weights and at all temperatures.*" In other words, that a grain of water, or a million of its atoms, when converted into vapor, will exercise the same "force and pressure" whether raised from a body of water at 32 or 212 degrees. By proceeding in this way, we may logically infer that the *temper-*

---

* For a full and exhaustive treatment of this subject, see CHAS. W. WILLIAMS' work on Heat, Water and Steam.

9

ature and pressure must be in the ratio of *quantity or number of atoms present* in a given space, *each representing unity* pressing, by the mere effect of accumulating numbers, the bulb of the thermometer, and producing a corresponding effect in the temperature.

When therefore, we say vapor (or steam) is at the temperature of 212°, it cannot be supposed that the temperature of each atom is 212°. Were that the case the aggregate of heat imparted would, by accumulation, become inconceivable. We are then compelled to consider the indicated temperature merely as *the sum of units then present*. By this simple process the alleged anomilies will be disposed of, while the *number* or *quantity present* in each *given space* will be the measure of *elasticity*, pressure, force, volume and temperature, and the several other conditions incident to steam in mass.

The misconception which so universally prevails comes from looking in the wrong direction for the results of the heat applied : —

1st. In assuming it to be *chemically* combined with the body of water, instead of with those of its elementary atoms with which it comes into contact; and 2d. Overlooking Dalton's well-established law that " vapors : *elastic fluids* " are but mechanically *mixed* with the liquid medium in which they may happen to be.

Neglecting these two important considerations, we look to the body of the water, *in its liquid form*, as the chemical recipient of the heat; in the very face of the *vapor* which we see mechanically disconnecting itself from the water, and thus carrying away that heat. Water, then, in the state of liquid, and at all temperatures, must be considered as a *mechanical compound* of liquid and vapor particles. So also the boiling point, and the temperature of 212°, must be considered without reference to the water, but rather as irrespective of its presence, as if it merely represented a vacuum.

SECTION VI.

EBULLITION AND CIRCULATION.

Having in a previous chapter described the phenomena accompanying vaporization, or the conversion of single atoms of liquid into those of vapor, and the peculiar repellent properties these atoms possess after such conversion, we will next consider the effect of a continued application of heat to water, and the process of EBULLITION AND CIRCULATION.

Ebullition has been so variously described by different writers that their accuracy may be reasonably questioned. There are certainly some points that will not bear the test of examination.

Says Mr. Williams: "In the first place, the *general mass* of a liquid does not boil." Ebullition, or boiling, is solely the local commotion which originates in the several groupings found at the bottom and then rising through the body of water. The bubbles (so often described) are the mere aggregates of· the vapor atoms *previously formed*, although invisible. Inasmuch as they must exist before rushing to the points or projections of rough surfaces or motes accidentally presented to them, — as may be seen by watching the process of ebullition — just as the vapor atoms in the air must have previously existed before they could be grouped by the electric action, and descend in a shower of rain drops, as seen after a thunder storm (and possibly for the same reason).

After quoting extensively from many writers on this subject, Mr. Williams says: "None of these writers refer to the true first cause or source of ebullition, namely, *the presence of vapor in the water in excess of saturation*, yet without this no groupings and, consequently, no ebullition can take place, as such does not begin until the quantity of vapor present approaches the point of saturation.

This point, as already stated, will be reached when the *diverging* or self-repellent force of the vapor atoms throughout the mass, are in equilibrium with the *converging pressure* arising from the density of the water medium surrounding them. Now this equilibrium, of itself, demonstrates the pressure and influence of the vapor, since if the heat had been absorbed by the water or liquid atoms, these atoms, having no repellent property and being influenced by gravity alone, would rise and remain uppermost: consequently, uniformity of temperature throughout the mass could not exist. This point of saturation, commonly called the boiling point, will then be determined solely by the quantity of vapor present in any given space: in water it takes place when that quantity indicates the temperature of 212° Fahr., in alcohol 176°, in sulphuric acid 630°, and in mercury 660°. Let it now be assumed that 1,000 atoms of vapor in any given space is the saturating quantity in water: until that quantity be present there will be no tendency towards these groupings and no ebullition, nor even then *unless there be some motes*, points or foreign matter present." As the rise in the thermometer above 212° indicates the presence of vapor in excess of saturation, this may be physically demonstrated by the discharge of such excess, either gradually or suddenly, by the introduction of a heavy foreign substance — any body that will fall to the bottom, such as pieces of coal, brick or iron, which serve as nuclei for groupings of the vapor.

It would be interesting to follow Mr. Williams through the great number of experiments profusely illustrated

in his work, by which he ascertained and fully demonstrated —

1st. The existence of the vapor in water.

2d. That the excess of such vapor beyond the
saturating point may be discharged 'gradually or suddenly.

3d. That ebullition is the mere result (mechanical or
electric) of the tendency of the vapor to rush into contact with any foreign matter that may be present and
furnishing points or nuclei for aggregation. These
small motes or points become nuclei, towards which the
vapor will rush so soon as the saturating quantity shall
be present, *but not a moment sooner*. These floating
objects will cause the appearance of a continuous stream
of aggregates or small globules.

The aggregation of these atoms (already existing in
steam) have been mistaken for its generation; and
ebullition is merely accidental and has no reference to
the generation of vapor and is solely the result of the
aggregation of *atoms previously* formed, such aggregates
being composed exclusively of such vapor atoms as are
in excess of the saturating quantity.

Ebullition being a fact, it must be in accordance with
some wise provision of nature for a useful purpose.
As it appears to have no direct influence in promoting
vaporization, what then may be its practical value or
effect in the economy of nature? At least two important
objects may be inferred: 1st. The prevention of a
useless, if not dangerous, accumulation of vapor in
liquids, under the accumulation of heat. 2d. The producing of that *all-important movement of circulation* —
the element of equal distribution of heat and vapor
throughout the mass.

As regards the *first*, vapor cannot escape from a
liquid, only at its surface : there would necessarily be an

ever-prevailing tendency to its *accumulation in the water*
were there no other means of effecting its discharge
than would be due to the area of that surface under the
mere operation of diffusion.

Now this object is directly effected by the rapid collection of the vapor atoms in the groupings which are seen
in ebullition. As each group is formed by reason of its
bulk and levity, it rises to the surface in the form of a
globule, and with an accelerated force escapes into the
air, the body of the liquid thus being relieved of its
presence. Of the second purpose: When these groups
and globules are produced, they rise with a force and
momentum due to their enlarged volume and levity.

These aggregates of the gaseous elements of vapor may,
in their effect upon circulation, be compared to that of a
balloon, mechanically forced upwards by the pressure
from beneath of the heavier particles of the air. We
know that the gas with which the balloon is filled, *if
liberated*, would be discharged into the air, each atom
ascending with a force due alone to its own specific
gravity, but which would necessarily be slow and comparatively ineffective. When, however, the myriads of
atoms of gas are brought together and confined within
the balloon envelope, the levity of the whole gives it an
ascentional force and rapidity which carries it to the
higher regions of the atmosphere. In the same way,
each group of globules or vapor, formed at the bottom
of a body of water, is productive of precisely similar
results. The *secondary* and equally important result is
that by which *circulation* is directly effected. As the
balloon ascends, and on each step of its progress
upwards, it would leave a vacuum *below* it (as a ship
moving through the water behind it), were it not
that the space is at once filled with the succeeding
portions of the air (or water), and a mechanical action is
produced.

Circulation, then, is the result of *quasi*-induced currents, consequent on the movement of a body through air or water, and in proportion to the rapidity of motion.

The inferences here drawn — from these investigations and experiments, to determine the progress and influence of heat on matter and its motions — do not rest on any baseless hypothesis, but on a clear view of the constant and unerring laws of nature as far as they appear to our view or within the range of our reason.

SECTION VII.

VAPOR IN WATER, AIR AND STEAM.

Very few, if any, writers on the subject of elastic fluids recognize the existence of *vapor in water* in its separate, independent character. It seems strange that the visible appearance of the great quantity that rises out of a body of so-called boiling or hot water should not have suggested the idea that it must, previous to its escape, have existed in the water; and that without such separate and independent existence its volume could not have been enlarged, diffusion or divergence would have been arrested, pressure nullified and elasticity itself have ceased to exist.

Atoms of vapor until affected by heat cannot be distinguished *in or* out of water, by reason of their minuteness; yet we have sufficient evidence to convince us of their presence in both cases. While they remain apart, with their several diverging properties, they remain invisible; as soon, however, as they converge or congregate and form *globules*, they come within the reach of our senses. This may take place both *in* or out of the *body of water*. When in the water they form globules, which, by reason of their greater levity, rise to the surface, then burst and pass into the air *above* it; out of the water in a similar way, when they come in contact with and are surrounded by a film of liquid particles (these forming a visible envelope, producing what is called vesicular vapor) and appear in a cloud.

If vapor exists in the water, it may be asked, why it does not rise at once to the surface and pass away into the air, by reason of its greater levity than water. It may be asked with equal reason, why the vapor, which exists in *the atmosphere near the earth*, does not rise at once to the *more rarefied* upper regions and leave the lower without any? The cause and the reason are alike in both cases, and are found in the nature of vapor *as an elastic fluid* filling the entire space. The air and the water are but mediums (as regards density and pressure) in reference to either the upper or lower regions of the atmosphere, or the still lower medium of water; the whole being but a question of degree, the vapor atoms being always in a state of mutual repulsion, without regard to the medium in which they may be formed. It is this diffusive action which prevents any permanent irregularity in quantity in any one portion, whether it be the atmospheric medium or a fluid of any kind.

"Homogeneous elastic fluids," says Prof. Dalton, "are constituted of particles that repel one another with a force decreasing as the distances of the centers of the particles." This law of repulsion and relative distances being general, it is equally applicable to *vapor* or *steam* as to any other *elastic fluid*. This is placed beyond all doubt in his precise statements, which in substance are as follows: 1st. That *vaporized bodies* cannot, on any scientific principle, be classed in a distinct category from *permanently elastic fluids*. 2d. That when two or more gases or vapors are put together, either into limited or unlimited space, they will finally be arranged each as if it occupied the whole space, and the others were not present. 3d. That they retain their elasticity or repulsive power *amongst their own particles* just the same *in the water as out of it*, the intervening spaces having no other influence in this respect than a mere vacuum.

"This," observes Mr. Williams, "is practically the most important feature of Dalton's great discovery of *diffusion*, whether in reference to meteorology or

physics — to temperature in the atmosphere or in the
water, to the properties peculiar to the liquid or the
vaporous state."

We conclude, then, that with reference to varying
degrees of indicative temperature, the quantity of vapor
generated in or *injected into* a body of water, be it great or
small, the repulsive power among its particles will cause
them so to diffuse themselves that no part of the liquid
mass shall be without its due proportion. Berthollet
demonstrated this law laid down by Dalton — this inter-
penetration or movement of gases, and called it diffusion;
and says: " In a mixture of gases, the pressure, or
elastic force, exercised by each of the gases is the same
as if it was alone." The question then arises, Is not this
law equally applicable to the *elastic fluid, vapor?* The
analogy is brought still closer by considering the gas
(vapor) *mixed with water.* He says : " When a gas comes
into contact with *a liquid,* the gas is absorbed in a
quantity varying with the pressure to which it is subjected." Thus the constituents of the atmosphere are
always formed in the water with which it is in contact;
" and for the same gas, the same liquid, and the same
temperature, the weight of gas absorbed is proportional
to the pressure: that is, that at all pressures the volume dissolved (mixed with it) is the same." Now,
pressure, or the effect of diffusion, being the same, what
is the amount of that pressure *in water* to which the
vapor is subjected? This can only be determined by
reference to the respective densities of the two media,
the water and the air. Here then is to be found the
true amount of effective pressure *from without,* to which
every gas or vapor *forced* into or *formed in* water must
be subject.

This is still more clearly illustrated by Professor Silli-
man when speaking of *molecular repulsion,* as follows :
" If a definite volume of air is admitted into a vacuum of

twice that capacity, it does not, like a solid or liquid body, retain its *original volume*, but expands and *fills the whole empty space*; and the same offers a resistance when these particles are brought together by mechanical pressure. A similar resistance to compression is displayed in *liquid* and solid bodies." Here it is shown that molecular repulsion is equally referable to that of vapor or other elastic fluid, namely : in the force which acts repulsively among their particles. On the elastic force of heat, he adds : " Since the *accumulation* of heat causes the atoms of bodies to *separate*, and its *removal* causes them to *approach each other*, it must be admitted that whatever may be the nature of heat, it acts *as a repulsive force.*" It is probable that this *repulsive force* cannot be explained on any other known principle than that of *electricity.* Bodies (atoms of vapor are bodies) in the same electrical state have a mutual repellent action. It is singular that this repulsive force is generally altogether ignored by writers when treating of *vapor in water*, although all admit it to be an elastic fluid and endowed with like properties. While all authorities admit the existence of vapor in *the air*, yet ignore it when in the denser medium of *water*, the prevailing theory would imply that it can exist in *no other proportion* than in the enlarged, expanded state, due to the pressure of the atmosphere.

Dalton observes : " Vapor exists at all times in the atmosphere, and is one and the same as steam or vapor at 212° or upwards." Another writer remarks : " This goes far to confirm the case of *unity* of heat (as already explained), seeing how infinitely minute must be the equivalent of heat in each atom of vapor while in the atmosphere; and that 212° is merely the result of a given quantity or number of such accumulated atoms then present in space. Practically, then, there would appear a greater difficulty in conceiving the existence of vapor in the *atmosphere* than in *water*, seeing that air

is a positive refrigerator and conductor of heat, whereas
we have no reliable grounds for saying that a liquid is
the *former*, and we know it is not the *latter*." This same
writer, quoting Dr. Henry ("This notion which gives
identity to vapor formed by heat in vacuum is ingen-
ious; but how the vapor should *ascend till it arrives at
the air* of the same density is not conceivable"), replies as
follows : "Certainly not, so long as the heat is assumed
to be absorbed *by the water* while still retaining its liquid
form, and that vapor cannot co-exist with water in a
separate and independent capacity. So soon, however,
as these errors are repudiated, the ascent of vapor
through the liquid mass will be as intelligible as the
ascent of a cork from the bottom to the surface in a
vessel of water. Here is the stumbling-block which has
so long stood in the way; yet, if we only look at the
formation of vapor, not in the mass, but with reference
to its atoms, separately and successively receiving their
quotas of equivalent units of heat, and obtaining their
distinct properties of levity, repulsion and elasticity, all
difficulties will be at an end." The last writer (Mr.
Williams) who so ably defends this theory—namely, that
vapor can and does co-exist with and in water in a
separate and independent capacity — appears to fully
demonstrate his proposition by a great number of
experiments in his work on "Vapor in Water" and
finally sums up in the following paragraph : —

"So far from looking to what are called 'high authorities' as safe, practical, experimental or scientific guides,
we have in their writings on this question but a melan-
choly array of contradictions and anomalies, from
which practical men can find nothing on which they can
rely, and are compelled to admit the necessity of experi-
menting and reasoning for ourselves if we would avoid
these discrepancies which embarrass both the subject
and the student."

SECTION VIII.

EVAPORATION — ESCAPE OF VAPOR.

Under the head of vaporization, the formation of *vapor* atoms from *liquid* atoms has been described. We will now consider the conditions under which these vapor atoms *are separated and escape from the water.*

But for the inaccuracy in the use of language — which implies inaccuracy in reasoning — the distinction between the *generation* of vapor and its *escape* into the air would be obvious. Yet among writers of the highest authority the terms *vaporization* and *evaporation* are frequently used as synonymous terms, and so confounded and misplaced as to lead to practical errors. Notwithstanding the indiscriminate and careless use of these terms, there are no two processes in nature more distinct : *vaporization* being the conversion of liquid atoms into those of *vapor* by the absorption of heat; *evaporation* being the "mere *escape* of the vapor atoms *so formed*, and without reference to heat." As an instance of the misapplication of these terms : "*Evaporation*," says the *Encyclopedia Britannica*, "is that process by which water and liquids are *converted into steam*, an elastic fluid, and dissipated in the atmosphere." Here the *cause* is confounded with the *effect*. Water certainly is not " converted into steam " by evaporation. The term has no meaning but as expressing the escape or removal of vapor, and there can be no evaporation until there be *vapor* to escape. With equal propriety it might be said that the process of generating gas in the retorts by heat, is the same as that by which it passes through the pipes and escapes from the street burners. The *generation* and *escape* in both cases are equally distinct processes. *Evaporation,*

then, is consequent of and subsequent to the previous
act of vaporization; a neglect of which distinction
involves the error of implying that liquid atoms are converted into vapor *only when* they *rise* and escape into
the air.

The Cyclopedia, in quoting the opinions of philosophers, upsets its own definition, as follows : "Aristotle
ascribed the *formation* of *vapor* to the action of fire."
"Halley supposes small spheres of water to be filled
with subtle fluid, so as to make them lighter than air."
Desaguliers asserts that water is capable of being converted by heat into an elastic fluid much lighter than
air. These authorities are correct in considering the
formation of *vapor* to be the union of liquid particles
with heat, but furnish no authority or intimation that
evaporation means the imparting of that heat. One
writer observes : "*Evaporation*, properly speaking, is
the result, or rather effect, of the intimate *union* of
elementary fire (heat) *with water*. By this union the
water and fire combined form an elastic fluid, specifically
lighter than air, and which is peculiarly distinguished by
the name of *vapor*." Here we have a correct description,
not of *evaporation*, but of *vaporization*, and it is only
when the vapor thus formed escapes that *evaporation*
begins.

All admit that "evaporation produces cold," which is
correct; and since, as increased temperature is derivable
from the increased quantity of vapor atoms present, so
the escape of these atoms is the same as the escape of
the heat which they (each) carried away, and by which
the sensation of cold is produced. Rees' *Cyclopedia*
says : "Cold is produced when any part of the human
body is moistened with water and the same is suffered
to evaporate." This is a direct case of evaporation : as
the atoms of such moisture must first be vaporized
before they can pass into the atmosphere.

By a very simple experiment, vaporization and subsequent evaporation can be demonstrated. Hold a champagne glass in the hand, then pour in cold water (colder the better); a sensation of cold is immediately experienced, indicating a loss of heat from the hand. What becomes of that heat? The common opinion would say, that it had gone to heat the water; but this is an error. It has gone, first to heat the glass, and then, not to heat the water, but to vaporize or convert the atoms of the water, in immediate contact with the glass, into vapor, each atom of the liquid receiving its unit of heat, and so becoming one of vapor. What is the proof of such conversion? Namely, the escape of that vapor and becoming visibly condensed on a glass or mirror laid on or over the top of the glass. Here we have: 1st, the vaporization of a portion of the water, and 2d, the escape of that vapor and its visible condensation, which is tangible and conclusive proof.

OF SPONTANEOUS EVAPORATION.

Unable to reconcile or account for phenomena of the escape of vapor at all temperatures, in the absence of any reliable explanation of place or means by which the vapor is formed, writers have recourse to the theory of " *Spontaneous Evaporation.*"

" Water slowly evaporates" (says Prof. Brande), "*under exposure to the air*; its vapor mixes with the surrounding atmosphere, and the process is usually called *spontaneous evaporation*: it takes place at all temperatures, and with a rapidity proportionate to the dryness of the air and the velocity of the current passing over it." " Here is an oversight of importance which merits attention " (says Mr. Williams). " The evaporation from water is here assumed to be in consequence of its *exposure to the air*. This is not the case; the

evaporation, or escape of the vapor is always going on,
whether the water be so exposed or confined in a vessel
apart from the atmosphere. If we half fill a large
bottle, and tightly cork it, we find the vapor continuously
rising from the water, and being condensed on the *upper*
part of the glass, always trickling down and again
returning to the water, so that no loss of weight is
sustained; while the *lower* part of the bottle, receiving
heat from the temperature of the room, produces a continuous supply of new vapor in the water." So there is
no necessity for this recourse to a supposed "spontaneous" action. The term seems to have been adopted as an
excuse, by writers unable to assign a sufficient cause for
the diminution of the mass of water exposed to atmospheric influence: it is, however, an unmeaning inapplica-
ble term when thus applied to matter.

All the movements in connection with evaporation
are subject to the immutable and well-known laws of
gravitation and diffusion; and if we examine the subject
carefully and experimentally, we shall find that the rise
and escape of vapor atoms (which is virtually evaporation) is as much under the influence of gravity as that
of a cork in water. The ordinary theory, however,
leaves us to infer that vapor atoms only have an existence,
as they rise from the surface of the water, and are then
and there formed without any further accession of heat.
In the light of the theory urged here, the expression
"*water evaporates*" is incorrect, unless we add, "*in the
form and state of vapor.*" This, however, involves the
distinction between *liquid* and *vapor* atoms, and still
leaves the question open: Where was the *vapor* before
its escape, and how was it formed previous to its going
off in evaporation? When this distinction is admitted, the
difficulties of coming to an intelligent conception of this
subject will be greatly diminished. The whole difficulty

arises from considering the rise of the vapor from the surface of the water, *apart from* its previous existence in it. Of the fact of the vapor rising out of the water there can be no more conclusive evidence than physical test of its condensation. In this there can be no deception, as our senses furnish absolute proof.

It is found (by experiment) that there is a continued harmony " between the loss of vapor and the loss of temperature, which indicates a corresponding harmony between the weight and heat of each evaporating atom. Thus, in harmony with this theory, we are enabled, by observation and logical deduction, to arrive at accurate results, establishing a positive law, namely: that as each atom of vapor has its equivalent unit of heat, an increase or diminution in number of the one, must have its correlative in a commensurate increase or diminution of the other."

" A further result is necessarily deduced from this harmony of *heat* and *quantity*, namely, that of *time*, as the escape of this surface stratum will contain a smaller number of atoms: the result being that a commensurately longer time will be required in producing given amounts of evaporation — that is, given reductions on the gross weight of the body of water. Thus we see how the reduction of temperature must be the mere co-efficient of the loss of vapor." Dalton made some accurate experiments on these points, his attention being directed to the question of *time* rather than the corresponding differences in *temperature* and *weight*.*

---

* " Water, freely exposed to the air, evaporates at all tempera-ures, even when in the state of snow or ice. The rapidity of evaporation is, however, much increased by *warmth*. Thus, in a calm atmosphere, Dalton found that when, from a certain surface, the evaporation from boiling water proceeded at the rate of 40 grains per minute, it was 20 grains at a temperature of 180°, 13 grains at 164°, 10 grains at 132°, and so on." — HERSCHEL, in *En-cyclopedia Brittanica.*

Numerous experiments have since been made, all establishing the fact that the loss of *weight* by evaporation, as it is the *cause*, so it must be commensurate with the loss of temperature — *time* alone being the varying incident in the process. Herschel says: "Evaporation never takes place without the abstraction of heat from the evaporating surface." A mere truism. We may with equal truth say we cannot remove any portion of a body of water without an abstraction of a commensurate weight: *heat* being as much an element of *vapor* as *weight* is of *water*. *Water* finally *does not* evaporate, but merely *parts with its vapor*. "In fact, the water — the liquid mass — has no direct action or influence on the process of evaporation, and is merely the denser medium in which, at the time, vapor happens to be; and which, acting the part of a vacuum, gives scope and capability for its diffusion. Hence evaporation is the escape of vapor, also heat, which is the element of its existence as vapor."

SECTION IX.

CONDENSATION.

" The term condensation is commonly applied to the conversion of vapor into water in the process of distillation. The way in which vapor commonly condenses is by the application of some *cold substance*. On *touching it* the vapor parts with its heat; and doing so, it immediately loses the proper characteristics of vapor and becomes water. If heat be withdrawn from steam or vapor, it no longer remains in the vaporous state, but resumes a liquid form. In this state it undergoes a great diminution of bulk, a large volume of steam forming only a few drops of liquid. Hence, the process by which the vapor passes from the aëriform to the liquid state, is called condensation." — *Encyclopedia Brittanica*.

The above quotation is a correct description of the cause, process and effects of condensation, or the reconversion of vapor into the liquid state. Applying this to the steam engine, the steam, owing to its elasticity, rushes into the condensor; but what " *cold substance* " does it there meet? This is the important point to consider. For the almost universally received theory is, that the steam or vapor meeting a body of *cold water*, imparts its heat to the latter, and is thereby instantly condensed, or reconverted into the liquid state. This erroneous, but prevailing theory, arises from the assumption that water, although a *non-conductor*, is nevertheless an *absorber* of heat, and overlooking the fact that water is not

a *substance* to which vapor can give out its heat, or,
which is the same thing, that heat is absolutely *absorbed*
by water. If water could convert vapor (as generally
assumed) into the liquid state by abstracting its heat,
the result would be that vapor could *not be formed*, or,
at least, would have no dynamic effect; for the moment
the first atom of the liquid was *converted* into one of
vapor, by the heat, it would as quickly be *reconverted by
the mass of water* surrounding it. No permanent condition could exist, and no body of vapor could be formed.
It is not possible to reject this logical inference.

What really does take place "when vapor is thrown
into what may be called an *atmosphere of water*, each
atom is at once compressed or reduced in influence and
prevented exercising its full expansive power by the
combined densities of the two media — the water and
air. No diminution, however, of the temperature of the
vapor atoms follows. They merely remain, with their
compressed volumes in the water, until they escape into
the atmosphere, or by contact with some *cold substance*
lose their heat, and are then *bona fide* reconverted into
liquid form." Water, then, or *any* liquid, is not a substance to which heat can be imparted; or, in other
words, heat cannot be received and *retained by liquid
particles*, each of which is susceptible to an instantaneous
change in its statical or electrical condition, by the accession of heat. It would be as reasonable to expect
that atoms of ice could receive or *absorb* heat, and
having their temperature raised *still retain their crystalized form* (or state of ice), as that those of water could
receive it and retain their STATUS *of a liquid.*

"Air is an elastic fluid," Mr. Williams observes,
"and is a recipient of heat, since its *status cannot be
altered* by it, there being no *fourth state* into which it
might enter by a further accession of heat. Besides,

being also a *conductor* of heat, it is capable of receiving and imparting it to others, from atom to atom. In this way, the vapor in the atmosphere when brought into contact with a body of colder air and more or less of the vapor atoms (according to the amount of atomic contact realized between them), gives out its heat to those of the air, returns to the liquid form, and produces the effect of visible clouds."

When we consider the extreme miscibility of elastic fluids, or aëriform matter, and the extent of surface for mutual contact presented by the aggregation of the millions of atoms which compose bodies of air and vapor, we can readily account for the rapid condensation of the vapor atoms in the atmosphere, when brought into connection or collision with currents of colder air. To these currents, then, may be attributed all atmospheric changes of temperature and humidity, from clouds, fogs and rain, up to the more rapid discharges accompanying electric disturbances.

When we look into the changes in the electric condition of these vapor atoms, on their losing the property of *repulsion* simultaneously with their loss of *heat*, and thereby becoming negatively electrified, we have the key to the intensity and great quantity of the electric fluid that is set at liberty. Looking at the subject from this point of view, we are led to conclude that the ordinary theory that cold water absorbs heat (in condensation), thus reducing the vapor to a liquid state, — " annihilating it as vapor," — is an error; and until the *true recipient of heat is determined*, we must remain in the dark, to a greater or less extent, in regard to the principle in which condensation is effected, and the *best* means of perfecting the process in the steam engine.

The rapidity with which vapor parts with its heat furnishes strong proof of the views here presented — namely, of vapors being a mere aggregate of atoms, each

of which has its *unit of heat in combination*, all being
capable of parting, *at once*, with their respective units;
for no matter how numerous these atoms may be, the
result would be equally instantaneous: hence the importance of the extended surface (or units of surface) for
contact.

In refutation of the theory of *water* condensation may
be mentioned the fact, that when steam is injected into
cold water (in a separate vessel), instead of being condensed or reconverted into water, it appears in the same
visible, cloud-like form, as when the vapor is originally
formed in the water. (See chapter on Vaporization.)

If in this case the cold water converts the steam into a
liquid state, how are we to account for its reappearance?
Why does not the mass of water at once cool down
(annihilate) or reconvert the first and succeeding atoms
of vapor as fast as they are introduced? Facts like
these, one would suppose, are enough to shake confi-
dence in this condensing or annihilating theory. The
fact that a small jet of steam discharged into a body of
water is capable of almost instantly raising the whole
temperature to 212°, shows conclusively that the process
is not of liquification of vapor, but, rather, of the saturation of the fluid medium with vaporous atoms.

A comparison of the ordinary theory of condensation
by water with Dalton's theory of diffusion would show
that they are opposite and contradictory. Dalton's statement, now so generally endorsed, namely, " That steam
is but dissipated and diffused through the water, as if it
were *a vacuum*, and being an *elastic fluid* it *retains* its
properties irrespective of such medium," seems to be in
harmony with the facts. If the common theory is right
(and Dalton is wrong), the steam would be at once
annihilated by contact with the water. If Dalton is
right, it would simply be diffused through the liquid
medium, the same as if that medium were a vacuum.

SECTION X.

ON THE VACUUM.

It appears, then, that according to the prevailing theory, steam, upon being brought into connection with a body of cold water, thereby becomes liquified or re-converted into water — in fact, is annihilated as steam, and that (as in the steam engine) the result of this annihilation necessarily produces a vacuum in the cylinder.* This theory (although endorsed and supported by the leading authorities from Watt, down), by the investigations and experiments of Mr. C. W. Williams, has been pretty thoroughly ventilated, and, as we think, proven to be entirely erroneous. On this point Mr. W. says: " It certainly is not a desirable task to be anyway instrumental in weakening so agreeable a reminiscence, or questioning anything coming either from Arago or Watt; but in search after truth, however, and in justification of science, as no respecter of persons, it becomes necessary not only to question the truth of this particular mode of *converting steam into water*, but to characterize it as a misconception, if not an absolute fallacy."

Whatever, then, may be the merit of the *separate vessel*, which is, in truth, the great element of Watt's success, the theory as regards *the action of cold water* in producing the vacuum, is altogether erroneous.

No doubt the abstraction of the steam from the cylinder naturally led to the conclusion that it had actually *parted with its heat*, and that it entered the cold water,

---

* Professor Rankine says, "The ordinary condenser is a steam and air-tight vessel of any convenient shape, in which the steam discharges from the cylinder by a constant shower of cold water."

and by condensation produced the desired vacuum. So
far, however, from the steam losing its heat or being
condensed into the state of water, is it not merely mechanically diffused (as already observed on the Dalton
theory) among its particles, in proportion to the respective quantities of each? The cold water being (by means
of the injector) dashed against or spread over the inner
surface of the metallic condenser, the latter, *becoming
cold*, acts the part of a true *surface* condenser, just the
same as if the water had been made to act against the
outside, as in the still. In this way, an absolute conversion of the steam into water is effective, and a vacuum
produced in exact proportion to the extent of the available surface and its reduced temperature; and doubtless
if the inside surface be sufficiently extended, and cold
enough, the entire steam may be condensed, and a *perfect*
vacuum formed. Practically, however, but a moderate
portion of the steam is so disposed of. " That water,"
says Mr. Williams, " whatever may be its temperature,
is incapable of reducing steam to its previous state and
bulk of water, is susceptible of direct proof." He then
goes on at great length and proves this by numerous
experiments, and adds, " It may then be broadly and
unequivocally stated that there is no other mode by
which steam can be condensed — that is, reconverted
into water — than by the abstraction of its heat by
metallic refrigeration, as is done in the still or a system
of metallic tubes" : That, in a word, there is none other
than *surface condensation*.

From the foregoing discussion, it is evident from the
numerous extracts made that due consideration has not
been given to the important distinction between vaporization and evaporation, and as to how and where this
" invisible vapor " is actually formed seem to remain
unsettled problems in chemistry. Why this is so may,

perhaps. be accounted for by a lack of disposition, time
to investigate for ourselves, or the habit of adopting the
views of others without question or inquiry. In closing
the subject under consideration, the following quotation
is very appropriate and to the point: " If we trace the
history of any science, we shall find it a record of *mistakes and* misconceptions : a narrative of misdirected
and often fruitless efforts. Yet, if amidst all these the
science has made progress, the struggle through which
it has passed — far from evincing that the human mind is
prone to error rather than to truth — furnishes a decisive
proof to the contrary and an illustration of the fact,
that in the actual condition of humanity, *mistakes* are the
necessary instruments by which truth is brought to light,
or at least indispensable conditions of the process."

APPENDIX.

ELECTRICITY.

The "conservation of energy" is an important law of physical science; by which is understood, that for any manifestation of force in matter, there is or has been an equivalent force expended commensurate with the work done or new force developed. This law is recognized in all electrical phenomena, and, without the previous expenditure of energy, no electrical effects are produced.

While we may not be able to determine the nature of electricity, we are forced to conclude that there is an inexhaustible supply of this subtle element in nature, always available for use; and with an understanding of the laws governing the same and requisite mechanical appliances, we are able to collect this wonderful energy, control and direct its movements to the accomplishment of definite results.

No electrical effect, however, can be produced except at the expense of some other form of energy.

The progressive movement of electrical energy along a certain path is technically termed the "current," and the current in electrical parlance is always associated with the idea of motion. Just how or why this subtle element follows a definite path or "conductor" we do not know, but with the suitable apparatus we do know that it may be generated and directed in its movements and used to do our work.

The application of electricity for lighting and as a motive force is now so common that any one in charge of a plant should at least understand the meaning of the terms used by electricians in describing and operating electrical works; therefore we furnish a brief description of the

TECHNICAL TERMS.

CURRENT, — In electrical parlance, is always associated with the idea of progressive motion in some form. It is a form of electricity by which its energy and effects are conducted or transmitted from one point to another.

ELECTRO - MOTIVE FORCE. — The attribute of electrical energy manifested in the force of the current from one point to another along a path or conductor, and may be used as a motive power for doing work.

POTENTIAL. — As ordinarily used, signifies the amount of change from the normal state, from one point in relation to another in the same system. It has a relative meaning and expresses the difference in electrical power or level of one point in relation to another.

VOLT. — The *measure* of the difference in electrical potential. When any two points in a circuit differ in potential, there is a tendency for them to return to a normal state or equal potential. This tendency is represented by electro-motive force, the unit of which is called a *volt*.

CONDUCTOR. — The medium or path through which electric energy or its effects may be transmitted from one point to another. *Conductivity* represents the *degree* or capacity which any substance or path provides for the passage of the electrical *current*. Silver, copper and metals generally are *good* conductors, over or through which the current flows easily; while rubber, glass, slate, etc., are *poor* conductors.

INSULATOR.— A very poor conductor, or non-conductor of electricity.

OHM. — The unit of measure of *resistance.* For example, one hundred feet of No. 20, B. & S. copper wire has a resistance of about one *ohm.*

AMPERE.— Unit of measure, of the rate of flow, or velocity of the electric current.

CIRCUIT.— The entire circle or path through which a current travels.

OPEN CIRCUIT.— When at any point there is an insulator over which but a small current can pass.

A SHORT CIRCUIT.— When the current is diverted from its course, a path being opened having lower resistance than the one intended for it to travel in. When such a short cut is provided, a large part of the current may be diverted from its regular path and may prove a source of danger, by heating and firing the conductor by which the current is diverted.

CLOSED CIRCUIT.— When a circuit, by the use of good conductors, is complete in every part. Cutting a wire over which the current is passing, so that there is only air between their ends, opens the circuit, as then but a small current of electricity can pass.

INDUCTION.— Effects manifested in a secondary wire without any metallic connection, when the current in the primary wire varies in strength, owing to the near presence of another current or wire conductor.

RESISTANCE.— The property of a body, the tendency of which is to interfere or resist the flow of the *current:* and through which more energy is required to force the current through. *Insulators* have great resistance.

HIGH RESISTANCE.—Any material which presents great obstruction to the passage of a current, may be said to have high resistance. In a circuit composed of obstructions, such as a series of lamps, the resistance is comparatively high.

LOW RESISTANCE.— Other things being equal, every additional wire for the current proportionally reduces the resistance. The more the resistance between any two points is reduced, the more current can be made to flow through the circuit, with a given E.M.F. If the resistance of one wire be 40 ohms, the arranging of another wire parallel with the first, of the same resistance in itself, will make the joint resistance of the two wires half that of one, or 20 ohms; if four wires, with a resistance of 40 ohms each, the resistance of the whole system of the four wires will be reduced to one-quarter of the one, or 10 ohms.

ILLUSTRATION.— If there are 5 lamps in a circuit, each offering a resistance of 4 ohms, requiring a pressure of 50 volts to force the current through one lamp : these lamps being connected in a series, would offer a mutual resistance of 5 times 4, or 20 ohms; and to force the current through these 20 ohms, would require an electric motive power five times as strong as the one lamp. or 250 volts.

MULTIPLE ARC.—Apparatus arranged to operate a series of arc lights and so regulated as to deliver a current of constant strength proportional to the resistance of the circuit, the resistance varying with the number of lamps in operation. The apparatus being connected in multiple arc, as a rule, is of low resistance : for the more paths or wires by which the current can go from one point to another, the lower the resistance of the whole circuit.

INCANDESCENT LAMP.— Consists mainly of a filament of carbon, or other refractory substance, enclosed in an air-tight glass globe. The carbon filament is supported within the glass bulb by two terminal wires. generally of platinum, and the carbon may be of various form and size, proportional to the current it is to carry and the amount of light it is designed to furnish. The wires extend through the glass, and on the outside means are provided for making electrical connection with the carbon filament within.

ARC LAMP.— Is an apparatus for producing light by means of the voltaic arc. It consists of a mechanical device for holding two carbon pencils in a vertical line, one above the other, a small distance apart. The lower carbon is usually fixed and the upper one movable, and so adjusted that by means of electro magnets its movement is controlled by the current circulating through the lamp, so that the points of the carbon are kept a small distance apart.

INCANDESCENT LIGHTING.— The lamps are usually connected in multiple arc or parallel paths. If one lamp. having a resistance of 100 ohms, requires an E.M.F. of 100 volts to force one ampere of current through it, then with five lamps connected, the resistance of the system will be reduced to $\frac{1}{5}$ that of the one, or 20 ohms. The net resistance being reduced to 20 ohms, the E.M.F. of 100 volts will force five times as much current over the circuit as in the first case, or 5 amperes. As each lamp requires one ampere, the total current going over the several wires is just enough for the five lamps. Hence for incandescent lighting, a dynamo will be required that will keep the current at a constant pressure, and supply more current as the resistance of the external circuit is lowered by the addition or opening of other lamps.

ALTERNATOR.—A mechanical device providing means for shifting the current or changing the alternating current of the arc lamp to incandescent lamps. An apparatus provided for running either arc or incandescent lamps on the same circuit with same dynamo.

PARTS OF THE DYNAMO.

DYNAMO.—A machine for converting mechanical energy into electrical energy, consisting essentially of magnets mounted on a solid base and an armature and other parts so arranged as to rotate rapidly between the magnets.

ARMATURE.— The central point wherein the mechanical energy is transmitted to the pulley or drum, and converted into electrical energy. It usually consists of coils of wire wound upon a suitable frame-work, composed of many pieces, insulated each from the other, and all mounted on a shaft, capable of rapid revolution.

MULTIPOLAR ARMATURE.— Is made up with several soft iron cores wound with insulated wire, in which the

COMMUTATOR — Is composed of copper strips thoroughly insulated and mounted on a

SPOOL, — Placed on the same shaft as the armature. These armatures present a broken surface and are generally used in so-called alternating machines.

RING ARMATURES.— Have an air space between the hub and ring, and their diameter usually exceeds their length. They require a device called a

SPIDER.— Hub and spokes of non-magnetic metal, used in connection with rings of insulated iron wire in making up an armature of the Gramme type.

DRUM ARMATURES.— Do not require a spider, and their length usually exceeds their diameter.

POLE PIECES, YOKES OR BACK STRAPS.— The parts directly facing the armature and the parts by which the cores are fastened together or to the frame of the machine.

CORES.— The frame work on which coils of wire are wound, usually made of soft iron disks on a shaft and insulated by some thin non-magnetic substance.

FIELD MAGNETS.— Large iron pieces, wound with insulated copper wire, between the ends of which the armature rotates.

BRUSHES.— Bundles of wires or strips of copper springs, in contact with and bearing diametrically on the opposite sides of commutator. They assist in conducting the current to the external circuit.

POLES OR FIELD OF FORCE.— The central point of a powerful attraction between the ends of magnets, as illustrated by placing iron or steel near these magnets. Field of force, the radius of this invisable power of attraction for metals, in which, if coils of wire are rotated rapidly, an electrical current is generated, which lasts only while the motion continues.

MAGNETIC FRICTION.— A peculiar friction, which acts like a brake on the armature, when in motion. By the continuous motion of the armature in overcoming this friction, the current is generated in the coils of wire. The essential work of the dynamo is to pull the armature around against this friction (or backward tendency) by means of the mechanical energy applied to the dynamo.

RELATION BETWEEN CURRENT RESISTANCE AND ELECTRO-MOTIVE FORCE.

Certain relations exist between the resistance of a circuit, the current flowing through it, and the electro-motive force that drives the current through the circuit. The resistance of a circuit and the E.M.F. under which the electric current is flowing, being known, the value of the current in amperes may be determined by dividing the number which expresses the E.M.F. (in volts), by the number that expresses the resistance (in ohms). Thus: the E.M.F. as between the points of a circuit (as found by a volt meter) being 80 volts, the resistance between the same points being 20 ohms — $\frac{20)80}{4}$ amperes. Again, the resistance of any wire or part of a circuit being known, to find the E.M.F. required to drive a certain strength through this wire, multiply the number expressing the current in amperes by the number expressing the resistance of wire in ohms, and we have the number of volts. Thus: Resistance 20 ohms
Amperes 4
$\overline{80}$ volts.

Again, to find the resistance of a part of a conductor between two points, knowing the electro-motive force, divide the volts by the number of amperes, which expresses the strength of current, and we have the resistance in ohms. Thus: 4)80 volts, E. M. F.
20 ohms, resistance.

The well-known formula by which these relations may be stated is: C. representing amperes; E, the number of volts; R. the number of ohms. Expressed. $c = \frac{E}{R}$. In these formulæ the number above the line is always to be divided by the number below the line.

For example : desiring to know the value of current in amperes : formula (1), $C=\frac{E}{R}$.

If we desire the value of E.M.F. in volts : formula (2), $E=C\times R$.

When the resistance is required in ohms : formula (3). $R=\frac{E}{C}$.

By the understanding of these formulæ, many practical questions may be easily solved. For example : it is desired to know the resistance in a group of incandescent lamps when hot, in multiple arc. An ammeter placed in the circuit shows a reading of 10 amperes. The voltmeter, with one terminal connected with the wire leading to the group and the other terminal connected with the wire leading from the group, shows a reading of 80 volts. Referring to formula (3) for finding resistance : $R=\frac{E}{C}$. E and C are known, R we wish to find : $R=\frac{80}{10}=8$. The resistance of the group when hot, then, is 8 ohms.

If there were 10 lamps in the group, and we want to know the resistance of one lamp hot : then, as the net resistance of the 10 lamps united in multiple arc would be $\frac{1}{10}$ that of one, the resistance of one lamp hot in this case is 10×8 or 80 ohms, and the resistance of the lamp cold would be about twice as much as when hot, or 160 ohms. To solve the same problem by direct measurements would involve the use of expensive instruments, perhaps not available.

Again : If 25 arc lamps, hung in a series, require an 8-ampere current, and the maker has given their resistance (when burning) as 4 ohms each, and the circuit in which they are hung is made of No. 6 B. & S. wire and is two miles long ; then (ignoring the resistance of the dynamo), we find — by reference to the following table— that two miles of No. 6 wire has a resistance of about 4 ohms (if the joints are as they should be). Therefore, 25 lamps, each 4 ohms, would be 100 ohms ; this, plus the resistance of the circuit, would be about 104

ohms. Now, referring to the formula for finding E.M.F.
(2), we know C and R; C×R or 8×104=832 E. That
is, the E.M.F. at the terminals of dynamo would be
about 832 volts. By direct methods special and costly
volt-meters would be required.

RESISTANCE AND SAFE CURRENT OF WIRES.

No. B. & S. WIRE.	Resistance, 1,000 ft. OHMS.	Safe Current. AMPERES.
1	.12	127
2	.15	101
3	.19	80
4	.24	63
5	.30	50
6	.39	40
7	.49	32
8	.61	25
9	.77	20
10	.98	15
12	1.5	10
14	2.5	6
16	4.	4
18	6.3	3
20	10.	2

These formulæ and table are useful in laying out
wire for incandescent lamps, etc.

Forcing the current over a line means overcoming
resistance. and as a certain amount of electro-motive
force is disposed of in this way, it becomes of considerable importance to so proportion the wires as to get the
least loss consistent with conductors of reasonable price.

A careful study of the above will familiarize the mind
with knowledge which may be usefully applied in electrical
works generally.

MISCELLANEOUS WEIGHTS AND MEASURES.

A point $= \frac{1}{72}$ of an inch.

A line $= 6$ points $= (\frac{1}{12}$ of an inch.$)$

A palm $= 3$ inches.

A hand $= 4$ inches.

A span $= 9$ inches.

A link $= 7.92$ inches.

A chain $= 100$ links $= 66$ feet $= 4$ rods.

A fathom $= 6$ feet.

A nautical mile $= 6,086$ feet and $\frac{7}{8}$, nearly.

A barrel of flour $= 196$ pounds.

A barrel of cement $= 300$ pounds.

A ton of anthracite coal (broken) $= 42$ cubic feet.

A ton of bituminous coal $= 47$ cubic feet.

A stone $= 14$ pounds.

A load of lime $= 32$ bushels.

A load of sand $= 36$ bushels.

A cable's length $= 126$ fathoms $= 720$ feet.

An acre $= 10$ square chains.

A load of bricks $= 500$ in number.

A cord of wood $= 128$ cubic feet.

A cord foot $= 4$ ft. long x 4 ft. high, 1 ft. wide.

A cord $= 8$ cord feet.

A load of unhewn timber $= 40$ cubic feet.

A load of squared timber $= 50$ cubic feet.

A load of inch boards $= 60$ square feet.

A load of 2-inch planks $= 300$ sq. feet.

A cubic foot of tallow $= 59$ pounds.

A hundred of nails $= 120$ in number.

A thousand of nails $= 1,200$ in number.

A bushel of sand $= 123$ pounds

A bushel of lime $= 85$ pounds.

A ton of coke $= 95$ cubic feet.

WEIGHT OF WATER.

Water.	Pounds.
1 cubic in..........	.03627
12 " "	.434
1 " ft. (salt)....	64.3
1 " " (fresh)..	62.5
1.8 " " " ..	112.
35.84 cub. ft. " ..	2240.
1 cylindrical in.....	.02842
12 " "	.341
1 " ft.....	49.10
2.282 " "	112.
45.64 " "	2240.

U. S. Gallons.	
1 U. S. gallon......	8.355
13.44 U. S. gallons.	112.
268.8 " " .	2240.

Imperial Gallons.

Imperial Gallons.	
1 Imperial gallon..	10.
11.2 " gallons.	112.
224 " " .	2240.

1 cubic ft. water=7.48052 U.S. gals.
1 cylindrical ft. water=6 U.S. gals.

NOTE.—The center of pressure of a body of water is at two-thirds the depth from the surface.

To find the pressure in pounds per sq. in. of a column of water, multiply the length of the column in feet by .434. Every foot elevation is considered (approximately) equal to one-half pound pressure per. sq. in.

Steam.

Steam is an elastic fluid, generated by the action of heat upon water.

Steam, when separated from the water from which it is generated, follows the law of all other gases, expanding 1,459 of its volume for each additional degree of heat, while the pressure remains the same.

The temperature of steam is equal to that of the water from which it is formed, and its elasticity is equal to the pressure under which it is formed.

Total heat of steam at 212° is 1,178.

HEAT UNITS IN WATER, BETWEEN 32° AND 212° F. AND WEIGHT OF WATER PER CUBIC FOOT.

Temperature.	Heat Units.	Weight, lbs. per cubic ft.	Temperature.	Heat Units.	Weight, lbs. per cubic ft.	Temperature.	Heat Units.	Weight, lbs. per cubic ft.
32°F	0.	62.42	123°F	91.16	61.68	168°F	136.44	60.81
35	3.	62.42	124	92.17	61.67	169	137.45	60.79
40	8.	62.42	125	93 17	61.65	170	138.45	60.77
45	13.	62.42	126	94.17	61.63	171	139.46	60.75
50	18.	62.41	127	95.18	61.61	172	140.47	60.73
52	20.	62.40	128	96.18	61.60	173	141.48	60.70
54	22.01	62.40	129	97.19	61.58	174	142.49	60.68
56	24.01	62.39	130	98.19	61.56	175	143.50	60.66
58	26.01	62.38	131	99.20	61.54	176	144.51	60.64
60	28.01	62.37	132	100.20	61.52	177	145.52	60.62
62	30.01	62.36	133	101.21	61.51	178	146.52	60.59
64	32.01	62.35	134	102.21	61.49	179	147.53	60.57
66	34.02	62.34	135	103.22	61.47	180	148.54	60.55
68	36.02	62.33	136	104.22	61.45	181	149.55	60.53
70	38.02	62.31	137	105.23	61.43	182	150.56	60.50
72	40.02	62.30	138	106.23	61.41	183	151.57	60.48
74	42.03	62.28	139	107.24	61.39	184	152.58	60.46
76	44.03	62.27	140	108.25	61.37	185	153.59	60.44
78	46.03	62.25	141	109.25	61.36	186	154.60	60.41
80	48.04	62.23	142	110.26	61.34	187	155.61	60.39
82	50.04	62.21	143	111.26	61.32	188	156.62	60.37
84	52.04	62.19	144	112.27	61.30	189	157.63	60.34
86	54·05	62.17	145	113.28	61.28	190	158.64	60.32
88	56.05	62.15	146	114.28	61.26	191	159.65	60.29
90	58.06	62.13	147	115.29	61.24	192	160.67	60.27
92	60.06	62.11	148	116.29	61.22	193	161.68	60.25
94	62.06	62.09	149	117.30	61.20	194	162.69	60.22
96	64.07	62.07	150	118.31	61.18	195	163.70	60.20
98	66.07	62.05	151	119.31	61.16	196	164.71	60.17
100	68.08	62.02	152	120.32	61.14	197	165.72	60.15
102	70.09	62.00	153	121.33	61.12	198	166.73	60.12
104	72.09	61.97	154	122.33	61.10	199	167.74	60 10
106	74.10	61.95	155	123.34	61.08	200	168.75	60.07
108	76.10	61.92	156	124.35	61.06	201	169.77	60.05
110	78.11	61.89	157	125.35	61.04	202	170.78	60.02
112	80.12	61.86	158	126.36	61.02	203	171.79	60.00
114	82.13	61.83	159	127.37	61.00	204	172.80	59.97
115	83.13	61.82	160	128.37	60.98	205	173.81	59.95
116	84.13	61·80	161	129.38	60.96	206	174.83	59.92
117	85.14	61.78	162	130.39	60.94	207	175.84	59.89
118	86.14	61.77	163	131.40	60.92	208	176.85	59.87
119	87.15	61.75	164	132.41	60.90	209	177.86	59.84
120	88.15	61.74	165	133.41	60.87	210	178.87	59.82
121	89.15	61.72	166	134.42	60.85	211	179.89	59.79
122	90.16	61.70	167	135.43	60.83	212	180.90	59.76

Total Heat Evolved by Combustibles, and their Equivalent Evaporative Power, with the Weight of Oxygen and Quantity of Air Chemically Consumed.

KIND OF COMBUSTIBLE.	Pounds of Oxygen Consumed.	Pounds of Air Consumed.	Cubic feet of Air at 60° Fahr.	Units of Heat per lb. of Combustible.	Equivalent Evaporative Power of 1 lb. of Combustible Atmosphere at 212° Fahr.
Hydrogen......................	8.0	34.8	457	60.032	64.2
Carbon, making Carb. Oxide..	1.33	5.8	76	4.452	4.61
" " Carbonic Acid..	2.66	11.6	152	14.500	15.0
Carbonic Oxide................	.57	2.48	33	4.325	4.48
Light Carbonated Hydrogen..	4.0	17.4	229	23.513	24.34
Bi-Carb'ted Hydrogen Ol. Gas	3.43	15.0	196	21.343	22.00
Sulphur.......................	1.00	4.35	57	4.032	4.17
Coal, average composition....	2.46	10.7	140	14.133	14.62
Coke, dessicated..............	2.50	10.9	143	13.550	14.02
Wood..........................	1.40	6.1	80	7.792	8.07
Peat..........................	1.75	7.6	100	9.951	10.30
Lignite.......................	2.03	8.85	116	11.678	12.10
Asphalt.......................	2.73	11.87	156	16.655	17.24
Straw, 12¾ % moisture.........	.98	4.26	56	5.196	5.56
Petroleum	4.12	17.93	235	27.531	28.50

Combustion of Fuel

Is the result of a chemical union of carbon and oxygen. Perfect results require about 2¾ lbs. of oxygen to 1 lb. of carbon, properly mixed.

The fireman should understand this law and aim to supply the necessary amount of oxygen and secure the proper mixture.

PROPORTIONS OF CYLINDRICAL TUBULAR BOILERS.

Horse-Power.	Diameter.	Length.		Tubes.			Thickness Shell.	Thickness Head.	Length Furnace.	Stack.	
				Number.	Diameter.	Length.				Diameter.	Height.
	In.	Ft.	In.		In.	Ft.	In.	In.	Ft.	In.	Ft.
15	36	8	11	30	3	8	$\frac{1}{4}$	$\frac{3}{8}$	3	18	26
20	36	10	11	30	3	10	$\frac{1}{4}$		$3\frac{1}{4}$	18	30
25	42	11		38	3	10	$\frac{9}{32}$		$3\frac{1}{4}$	20	30
30	42	13		38	3	12	$\frac{9}{32}$		4	20	36
35	44	13		46	3	12	$\frac{5}{16}$		4	22	36
40	48	13	2	52	3	12	$\frac{5}{16}$		4	24	36
45	50	14	2	52	3	13	$\frac{5}{16}$		4	24	36
50	54	13	2	58	3	12	$\frac{5}{16}$		4	26	36
60	54	16	2	58	3	15	$\frac{5}{16}$		$4\frac{1}{2}$	26	45
70	60	16	4	76	3	14	$\frac{11}{32}$	$\frac{7}{16}$	$4\frac{1}{2}$	28	45
75	60	16	4	76	3	15	$\frac{11}{32}$	$\frac{7}{16}$	$4\frac{1}{2}$	28	50
80	60	17	4	76	3	16	$\frac{11}{32}$	$\frac{7}{16}$	5	28	55
90	66	16	5	100	3	15	$\frac{3}{8}$	$\frac{7}{16}$	5	32	55
100	66	17	5	100	3	16	$\frac{3}{8}$	$\frac{7}{16}$	5	32	55
125	72	17	6	132	3	16	$\frac{7}{16}$	$\frac{1}{2}$	5	36	60

The Construction of Boilers

Varies with conditions. The plain cylindrical and the flue boiler are used when the cost of fuel is low and the feed water is poor, with limited opportunities for repairs. The multitubular boiler is more complicated, but it is more economical and is the kind in general use. In the return tubular boiler, 15 square feet of heating surface is usually allowed for each horse-power. The ratio of heating to grate surface with stationary or "dead" bars is about 40 heating surface to 1 grate surface; with a good shaking grate, about 50 to 1 of grate surface.

WROUGHT IRON WELDED PIPE FOR STEAM, GAS, WATER OR OIL.

1 inch and below, butt-welded; proved to 300 pounds per square inch, hydraulic pressure.
1¼ inch and above, lap-welded; proved to 500 pounds per square inch, hydraulic pressure.

TABLE OF STANDARD SIZES.

Inside Diam. In.	Actual Outside Diam. In.	External Circumference. In.	Length of Pipe, per Sq. Ft. of Outside Surface. Ft.	Internal Area. In.	External Area. In.	Length of Pipe, containing one Cubic Foot. Ft.	Weight per Foot of Length. Lbs.	No. of Threads per In. of Screw.
1/8	.405	1.272	9.44	.0572	.129	2500.	.243	27
1/4	.54	1.696	7.075	.1041	.229	1385.	.422	18
3/8	.675	2.121	5.657	.1916	.358	751.5	.561	18
1/2	.85	2.652	4.502	.3048	.554	472.4	.845	14
3/4	1.05	3.299	3.637	.5333	.866	270.	1.126	14
1	1.315	4.134	2.903	.8627	1.357	166.9	1.670	11½
1¼	1.66	5.215	2.303	1.496	3.164	96.25	2.258	11½
1½	1.9	5.969	2.01	2.038	3.835	70.65	2.694	11½
2	2.375	7.461	1.611	3.355	4.430	43.36	3.667	11½
2½	2.875	9.032	1.328	4.783	6.491	30.11	5.773	8
3	3.5	10.996	1.091	7.388	9.621	19.49	7.547	8
3½	4.	12.566	.955	9.837	12.566	14.56	9.055	8
4	4.5	14.137	.849	12.730	15.904	11.31	10.728	8
4½	5.	15.708	.765	15.939	19.935	9.03	12.492	8
5	5.563	17.475	.629	19.990	24.299	7.20	14.564	8
6	6.625	20.813	.577	28.889	34.471	4.98	18.767	8
7	7.625	23.954	.505	38.737	45.663	3.72	23.410	8
8	8.625	27.096	.444	50.039	58.426	2.88	28.348	8
9	9.688	30.433	.394	63.633	73.715	2.26	34.677	8
10	10.75	33.772	.355	78.838	90.792	1.80	40.641	8

PROPERTIES OF SATURATED STEAM.

Pressure per steam gauge.	Temperature in Fahrenheit Thermometer.	Heat required to raise one pound of water from 32° to temperature of evaporation.	Latent heat in one pound of steam.	Total Heat in one pound of steam.	Density or weight of one cubic foot of steam.	Volume of one pound of steam.	Cubic feet of steam from one cubic foot of water.
Lbs. per sq. in.	Degrees.	Heat Units.	Heat Units.	Heat Units.	Lbs.	Cubic Ft.	Cubic Ft.
0	212.0	180.9	965.7	1146.6	.03797	26.336	1642
5	227.2	196.3	955.0	1151.2	.05	20.	1246
10	239.4	208.7	946.3	1154.9	.0619	16.16	1008
15	249.8	219.2	938.9	1158.1	.0736	13.59	847
20	258.8	228.4	932.5	1160.9	.0852	11.74	732
25	266.8	236.6	926.8	1163.3	.0967	10.34	645
30	274.0	243.9	921.6	1165.5	.1081	9.27	577
35	280.6	250.7	916.9	1167.5	.1195	8.37	521
40	286.7	256.9	912.5	1169.4	.1308	7.65	477
45	292.4	262.7	908.4	1171.1	.142	7.04	439
50	297.7	268.2	904.6	1172.7	.1531	6.53	407
55	302.6	273.2	901.1	1174.2	.1643	6.09	380
60	307.3	278.0	897.7	1175.7	.1753	5.70	336
65	311.8	282.6	894.4	1177.0	.1863	5.37	335
70	316.0	286.9	891.4	1178.3	.1973	5.07	316
75	320.0	291.1	888.4	1179.5	.2082	4.80	299
80	323.9	295.1	885.6	1180.7	.2192	4.56	282
85	327.6	298.9	883.0	1181.9	.23	4.35	271
90	331.2	302.6	880.4	1182.9	.2409	4.15	259
95	334.6	306.1	877.9	1184.0	.2517	3.97	248
100	337.9	309.5	875.5	1185.0	.2625	3.81	238

PERCENTAGE OF SAVING OF FUEL BY HEATING FEED WATER. (Steam at 60 lbs.)

INITIAL TEMPERATURE OF WATER.

Temperature when fed to boiler.	32°	40°	50°	60°	70°	80°	90°	100°	120°	140°	160°	180°
60°	2.39	1.71	0.86									
80°	4.09	3.43	2.59	1.74	0.88							
100°	5.79	5.14	4.32	3.49	2.64	1.77	0.90					
120°	7.50	6.85	6.05	5.23	4.40	3.55	2.68	1.80				
140°	9.20	8.57	7.77	6.97	6.15	5.32	4.47	3.61	1.84			
160°	10.90	10.28	9.50	8.72	7.91	7.09	6.26	5.42	3.67	1.87		
180°	12.60	12.00	11.23	10.46	9.68	8.87	8.06	7.23	5.52	3.75	1.91	
200°	14.30	13.71	13.00	12.20	11.43	10.65	9.85	9.03	7.36	5.62	3.82	1.96
220°	16.00	15.42	14.70	14.00	13.19	12.33	11.64	10.84	9.20	7.50	5.73	3.95
240°	17.79	17.13	16.42	15.69	14.96	14.20	13.43	12.65	11.05	9.37	7.64	5.90
260°	19.40	18.85	18.15	17.44	16.71	15.97	15.22	14.45	11.88	11.24	9.56	7.86

For practical purposes it is safe to estimate that for every 11 degrees of heat gained by the use of the heater there will be a saving of one (1) per cent in fuel. It is estimated that a square foot of exposed surface of a boiler will condense an amount of steam equal to one-third of a horse-power per hour.

DIAMETER, CIRCUMFERENCE AND AREA OF CIRCLES.

Diameter.	Area.	Circumference.	Diam.	Area.	Circumference.
In.	In.	In.	In.	In.	In.
1/8	.012	.3926	12	113.69	37.69
1/4	.049	.7854	12½	122.71	39.27
3/8	.110	1.178	13	132.73	40.84
1/2	.196	1.570	13½	143.13	42.41
5/8	.307	1.963	14	153.93	43.98
3/4	.442	2.356	14½	165.13	45.55
7/8	.601	2.748	15	176.71	47.12
1	.785	3.141	16	201.06	50.26
1¼	1.227	3.927	17	226.98	53.40
1½	1.767	4.712	18	254.46	56.54
1¾	2.405	5.497	19	283.52	59.69
2	3.14	6.283	20	314.16	62.83
2¼	3.97	7.068	21	346.36	65.97
2½	4.90	7.854	22	380.13	69.11
2¾	5.03	8.639	23	415.47	72.25
3	7.06	9.424	24	452.39	75.39
3¼	8.29	10.21	30	706.86	94.24
3½	9.62	10.99	36	1016.88	113.0
3¾	11.04	11.78	42	1385.4	131.9
4	12.56	12.56	48	1809.6	150.7
4½	15.90	14.13	50	1963.5	157.0
5	19.63	15.70	52	2123.7	163.3
5½	23.75	17.27	54	2290.2	169.6
6	28.27	18.84	55	2375.8	172.7
6½	33.18	20.42	56	2463.0	175.9
7	38.48	21.99	60		188.4
7½	44.17	23.56	62		194.7
8	50.26	25.13	64		201.0
8½	56.74	26.70	72		226.9
9	63.61	28.27			
9½	70.88	29.84			
10	78.54	31.41			
10½	86.59	32.98			
11	95.03	34.55			
11½	103.86	36.12			

To Determine Height of Chimney.

The area being known, also the rate of combustion, multiply the number of pounds of coal consumed under the boiler per hour by 12 and divide the product by the sectional area of the chimney in square inches; square the quotient thus obtained, which will give the proper height of the chimney in feet.

WEIGHT AND VOLUME OF CAST IRON AND LEAD BALLS
From 1 to 20 in.

Diam.	Volume.	Cast Iron.	Lead.
In.	Cubic In.	Lbs.	Lbs.
1	.5235	.1365	.2147
1½	1.7671	.4607	.7248
2	4.1887	1.092	1.718
2½	8.1812	2.1328	3.3554
3	14.1371	3.6855	5.7982
3½	22.4492	5.8525	9.2073
4	33.5103	8.7361	13.744
4½	47.7129	12.4387	19.569
5	65.4498	17.0628	26.843
5½	87.1137	22.7206	35.729
6	113.0973	29.4845	46.385
6½	143.7932	37.4528	58.976
7	179.5943	46.8203	73.659
7½	220.8932	57.587	90.598
8	268.0825	69.8892	109.952
8½	321.555	83.8396	131.883
9	381.7034	99.5103	156.553
9½	448.9204	117.0338	184.121
10	523.5987	136.5025	214.749
11	696.9098	181.7648	285.832
12	904.7784	235.8763	371.806
13	1150.346	299.623	471.806
14	1436.754	374.5629	589.273
15	1767.145	460.6959	724.781
16	2144.66	559.1142	879.616
17	2572.44	670.7168	1055.066
18	3053.627	796.0825	1252.422
19	3591.363	936.2708	1472.97
20	4188.79	1092.02	1717.995

To Find the Weight of Safety Valve

Required to balance a given pressure at a given distance from the fulcrum : —

Multiply the area of the valve by the pressure, and from the product subtract the weight of the valve and lever. Multiply the remainder by the distance of the stem from fulcrum and divide by distance of ball from fulcrum ; the quotient will be the required weight in pounds.

| | WEIGHT OF IRON. | | | PLATE IRON. |
Size.	Round Iron.	Square Iron.	Thickness.	Weight per Square Foot.
In.	Lbs.	Lbs.	In.	Lbs.
3/16	.118	.117	1/16	2.55
1/4	.163	.208	1/8	5.03
5/16	.257	.325	3/16	7.56
3/8	.368	.468	1/4	10.07
7/16	.501	.638	5/16	12.59
1/2	.654	.838	3/8	15.11
9/16	.833	1.01	7/16	17.62
5/8	1.02	1.30	1/2	20.14
3/4	1.47	1.87	9/16	22.66
7/8	2.00	2.55	5/8	25.18
1	2.61	3.33	11/16	27.69
1 1/8	3.31	4.21	3/4	30.21
1 1/4	4.09	5.20	13/16	32.73
1 3/8	5.94	6.30	7/8	35.25
1 1/2	6.89	7.50	15/16	37.76
1 5/8	7.91	8.80	1	40.28
1 3/4	8.01	10.20	1 1/8	45.32
1 7/8	9.02	11.71	1 1/4	50.35
2	10.47	13.33	1 3/8	55.39
2 1/8	11.82	15.05	1 1/2	60.42
2 1/4	13.25	16.87	1 5/8	65.46
2 3/8	14.76	18.80	1 3/4	70.49
2 1/2	16.36	20.80	1 7/8	75.53
2 5/8	18.03	22.96	2	80.56
2 3/4	19.79	25.20		
2 7/8	21.63	27.55		
3	23.56	30.00		

Strength of Boiler Plates.

The tensile strength of American boiler iron is 40,000 to 60,000 lbs. per square inch. Very high tensile strength in boiler iron is apt to lack homogeneousness and toughness. Toughness of boiler plate iron better stands irregular strains and shocks.

www.ingramcontent.com/pod-product-compliance
Lightning Source LLC
Chambersburg PA
CBHW021806190326
41518CB00007B/466